Springer Series in Synergetics

Editor: Hermann Haken

Synergetics, an interdisciplinary field of research, is concerned with the cooperation of individual parts of a system that produces macroscopic spatial, temporal or functional structures. It deals with deterministic as well as stochastic processes.

A.S. Mikhailov

Foundations of Synergetics I

Distributed Active Systems

With 68 Figures

Springer-Verlag

Berlin Heidelberg New York London
Paris Tokyo Hong Kong Barcelona

Professor Alexander S. Mikhailov
Department of Physics, Moscow State University,
SU-117234 Moscow, USSR

Series Editor:
Professor Dr. Dr. h. c. Hermann Haken
Institut für Theoretische Physik und Synergetik der Universitat Stuttgart,
Pfaffenwaldring 57/IV, D-7000 Stuttgart 80, Fed. Rep. of Germany and
Center for Complex Systems, Florida Atlantic University,
Boca Raton, FL 33431, USA

ISBN-13:978-3-642-78558-0   e-ISBN-13:978-3-642-78556-6
DOI: 10.1007/978-3-642-78556-6

Library of Congress Cataloging-in-Publication Data. Mikhailov, A. S. (Alexander S.), 1950– Foundations of synergetics / A. S. Mikhailov. p. cm.– (Springer series in synergetics ; v. 51) Includes bibliographical references and
index. Contents: 1. Distributed active systems.   1. System theory. 2.
Self-organizing systems. 3. Chaotic behavior in systems. I. Title. II. Series: Springer series in synergetics ; v. 51,
etc. Q295.M53 1990 003'.7–dc20 90-10102

© Springer-Verlag Berlin Heidelberg 1990
Softcover reprint of the hardcover 2nd edition 1990

2154/3140-543210 – Printed on acid-free paper

*to the memory
of my teachers*

Preface

This book gives an introduction to the mathematical theory of cooperative behavior in active systems of various origins, both natural and artificial. It is based on a lecture course in synergetics which I held for almost ten years at the University of Moscow. The first volume deals mainly with the problems of pattern formation and the properties of self-organized regular patterns in distributed active systems. It also contains a discussion of distributed analog information processing which is based on the cooperative dynamics of active systems. The second volume is devoted to the stochastic aspects of self-organization and the properties of self-established chaos.

I have tried to avoid delving into particular applications. The primary intention is to present general mathematical models that describe the principal kinds of cooperative behavior in distributed active systems. Simple examples, ranging from chemical physics to economics, serve only as illustrations of the typical context in which a particular model can apply.

The manner of exposition is more in the tradition of theoretical physics than of mathematics: Elaborate formal proofs and rigorous estimates are often replaced in the text by arguments based on an intuitive understanding of the relevant models.

Because of the interdisciplinary nature of this book, its readers might well come from very diverse fields of endeavor. It was therefore desirable to minimize the required preliminary knowledge. Generally, a standard university course in differential calculus and linear algebra is sufficient.

The list of references is not complete, since it is mainly intended to provide suggestions for further reading. Occasionally, I give more detailed references to papers by Russian authors that may be less known outside this country.

The first impetus for my studies and lectures came from the works of Prof. H. Haken about twenty years ago. I consider it an honour that he recently suggested the publication of this book in the Springer Series in Synergetics. I must also acknowledge very stimulating discussions with Prof. Haken during my stay in Stuttgart.

Having completed this work I feel it appropriate to emphasize how much I am indebted to the late Prof. A.N. Kolmogorov, Prof. I.M. Lifshitz and Prof. Ya.B. Zeldovich who taught me, at different times, their approach to mathematics and theoretical physics.

I would like to thank Prof. D.S. Chernavskii, Prof. W. Ebeling, Prof. Yu.L. Klimontovich, Prof. V.I. Krinskii and Prof. J.J. Tyson for discussions that clarified some important issues. I express my special and warmest thanks to my friends and coworkers Prof. V.A. Davydov, Dr. N.A. Sveshnikov and Dr. V.S. Zykov. Finally, I am very grateful to my numerous students who, by posing sharp questions and making fresh comments, have greatly facilitated my task.

Moscow, June 1990   *A. Mikhailov*

Contents

1. Introduction

If you look in the *Concise Oxford Dictionary* of current English, you will find that the word "synergetics" does not appear there. However, the dictionary does contain the term *synergy* which is interpreted as "a combined effect of drugs, organs, etc. that exceeds the sum of their individual effects" and is derived from the Greek *synergos*, working together. Synergetics, as a separate interdisciplinary research direction, was introduced only about two decades ago by H. Haken [1.1, 2] who had noticed profound similarities between the fundamental mathematical models which are used to describe the cooperative behavior of active systems in physics, chemistry, biology, and social sciences.

For a long time, and until quite recently, many people believed in the possibility of constructing a universal theory where the models of a higher level would be systematically derived from the more fundamental ones, starting presumably with the laws of physics. This so-called *reductionist* approach is now considered to be naive and unrealistic. Experience shows us that, although there is often a certain degree of consistency between the related fields, mathematical models are very rarely derived from one another. As a rule they are proposed in a phenomenological way, capturing the most essential features of the observed behavior.

To apply the phenomenological approach, it is desirable to keep a stock of simple mathematical models of collective behavior that are fairly well understood. A specialist can, when occasion demands, browse through this stock and choose the model that is most appropriate in his or her particular case.

Usually, the available model does not describe all significant features of the observed phenomenon and a further refinement or fitting is required. This can be done by an interested specialist. However, a thorough analysis of generic models is best performed by a professional mathematician (or theoretical physisist) who might also suggest some useful modifications or extensions.

Synergetics can be considered as a shop where fundamental models of cooperative behavior are worked up and their careful mathematical analysis is performed.

The generic models of cooperative behavior deal with fairly *uniform* systems made of *simple* elements. At a first glance, this condition is not representative of biological or social systems. Even a single cell represents an extremely complicated object, not to mention the extreme complexity of human beings who are the individual elements of social systems. But closer examination reveals that in cooperative interactions these elements often act as if they *were* simple units that can be described by a set of a few variables. Their vast internal complexity is not directly manifested in their interactions (or it leads only to some minor variations which are effectively washed out by statistical averaging).

This property of living systems is not accidental: If we were to allow the elements to reflect all their internal complexity in the interactions, then the system as a whole would most probably not be able to display any stable and predictable behavior. Since a purposeful existence and operation usually implies predictability, systems with this property are naturally selected in the process of evolution.

As was noted by *von Bertalanffy* [1.3] and recently elaborated by *Corning* [1.4], living systems represent *hierarchies* of self-organized subsystems. At each level of the hierarchy, we have a system of sufficiently autonomous units that interact in a simple manner one with another. Their interactions create a coherent pattern which in turn plays the role of an element of the next level. Only weak influences are exerted by higher-level units to control the lower-level subsystems.

Thus the impressive variety that we find in the activities of living systems is produced by a combination and interplay of many simple ordered patterns that correspond to different structural levels. Ultimately, it is this notion that makes a mathematical theory of living systems feasible.

From this point of view, the self-organization phenomena which are observed in some inorganic (physical or chemical) systems (see, for instance, [1.5, 6]), differ from biological self-organization not in the types of individual patterns, but rather in the much lower complexity of the emerging hierarchical structures.

As was mentioned above, synergetics studies the cooperative behavior of *active* systems. In the common sense of this word, activity means a certain degree of autonomy from the environment that allows an element to perform certain transitions or motions which are not directly controlled from outside. All living systems, biological and social, possess this property almost by the definition. However, in chemical or physical systems this property is not common. The autonomous activity of such systems is possible only under certain special conditions.

The laws of statistical physics and thermodynamics tell us that, in general, a physical system tends towards a state of thermal equilibrium in which activity ceases. The only exception is observed in *open* systems that receive a flux of energy from some external source and dissipate it further into the environment. While the energy is flowing through the system it permanently maintains a non-equilibrium state, thus permitting continuous activity.

All living cells and biological macro-organisms have such permanent energy input in the form of food (or light, in the case of plants). The importance of energy flux for living creatures was first stressed by *von Bertalanffy* [1.3] and *Schroedinger* [1.7]. The theory of self-organization in open systems was largely developed by *Prigogine* and coworkers [1.5, 8].

In the next section we formulate the basic mathematical models of distributed active systems which will be thoroughly investigated in subsequent chapters. The second section is devoted to a discussion of theoretical questions related to the possibility of constructing *artificial "living systems"*, including their applications in the field of information processing.

1.1 Basic Models of Distributed Active Systems

Generally, a distributed active system can be viewed as a set of interacting active elements, which are either distinct units or represent sufficiently small pieces of a continuous medium. The simplest way to model an individual active element is to picture it as a *discrete automaton.*

Suppose that, by observing the behavior of an active element, we determine its few qualitatively distinct states or functional regimes. If the transitions between such steady states or regimes are fast enough, the transient processes can be neglected and we can assume that at any given moment this element stays in one of the few discrete states. Enumerating these states by integers $a = 1, 2, \ldots, K$, we can specify the state of such element at any time moment t by a certain value of a.

An automaton can change its state either after prescribed regular time intervals or in a probabilistic manner at irregular times. In the first case it suffices to consider an element only at regular times t_n before each transition, i.e. to record the states $a^{(n)} = a(t_n)$ at these times.

If an automaton is *deterministic*, its next state $a^{(n+1)}$ is determined by the sequence of previous states $a^{(n)}$, $a^{(n-1)}$, $a^{(n-2)}$, $\ldots$. If, furthermore, the automaton has no memory, its next state $a^{(n+1)}$ is completely determined by the state $a^{(n)}$ at the present time, and is independent of the history of transitions, i.e. we have

$$a^{(n+1)} = F\left(a^{(n)}\right) , \tag{1.1.1}$$

where $F(a)$ is some integer function that maps any possible state of an element into some other (or the same) one.

For a *probabilistic* automaton the next state is not uniquely defined. Instead, only the probability of transitions is specified. Below we shall not further discuss the properties of probabilistic automata; such a discussion is reserved for the second volume.

In some situations an element changes its states at regular times t_n, but its possible states do not form a finite set, since the state variable $a^{(n)}$ can take a continuous range of values. Sometimes the instantaneous state of an element cannot be specified by a single state variable and we then have to use several such variables $a_1^{(n)}$, $a_2^{(n)}$, $a_{(3)}^n$, $\ldots$, etc. In these cases it is convenient to construct a formal vector $a^{(n)}$ which has $a_1^{(n)}$, $a_2^{(n)}$, $a_3^{(n)}$, $\ldots$ as its components. Equation (1.1.1) is then replaced by

$$\boldsymbol{a}^{(n+1)} = \boldsymbol{F}(\boldsymbol{a}^{(n)}) , \tag{1.1.2}$$

where $\boldsymbol{F}(\boldsymbol{a})$ is a vector function of the vector argument, i.e. represents a set of component functions $F_1(a_1, a_2, \ldots)$, $F_2(a_1, a_2, \ldots)$, $F_3(a_1, a_2, \ldots)$, etc.

Proceeding still further, consider a situation in which the change of any state variable at any time step $\Delta t = t_{n+t} - t_n$ is very small, so that we can approximately write

$$a(t + \Delta t) = a(t) + f\big(a(t)\big)\Delta t . \tag{1.1.3}$$

In the limit of continuous time we then arrive at a *differential evolution equation*

$$\dot{a} = f(a) \,, \tag{1.1.4}$$

where the dot denotes the time derivative.

When we have several state variables a_1, a_2, a_3, ... , the single differential equation (1.1.4) will be replaced by a system of first order differential equations that can be written in the formal vector notations as

$$\dot{a} = f(a) \,. \tag{1.1.5}$$

Note that even if we model an active element by a system of differential evolution equations this is not equivalent to its complete description. In the majority of cases such elements have such a complicated internal structure that an immense number of variables would be required to provide a full description. If we try to incorporate all these variables into our model, which is designed to take into account interactions between the elements, this would certainly ruin the model. Hence, a careful preliminary analysis is usually required, which is aimed at finding out the genuine state variables that are important for interactions and that effectively characterize the momentary state of an element. Other element properties are either irrelevant for interactions or adjust adiabatically to the slow variation of principal state variables.

These slowly varying principal variables, which are responsible for interactions that result in an ordered pattern of activity in a cooperative system, are sometimes called the *order parameters*. Then the remaining variables, which follow the evolution of the order parameters, are viewed as *enslaved*. This terminology originates in the analogy with second order phase transitions in equilibrium physical systems that was elaborated by *Haken* [1.1, 9].

Now we can take into account interactions between active elements, beginning with the simplest case of discrete automata.

Suppose that we have a network of N such elements that are enumerated by an integer $j = 1, 2, \ldots , N$. The network is defined by indication of the links (or *connections*) between the elements. All the elements that are connected by a single link g to a given element j represent its nearest neighbors. Let us denote the set of the nearest neighbors of an element j as $O(j)$: if $j' \in O(j)$ then the element j' is a neighbor of j.

Generally, the next state of an individual automaton in the network is determined by the present states of this automaton and of its neighbors, i. e.[1]

$$a_j^{(n+1)} = G\left(a_j^{(n)}, \{a_{j'}^{(n)}\}\right) \tag{1.1.6}$$

where $j' \in O(j)$.

If the network is not homogeneous, the law of mapping can vary from one element to another, so that G in (1.1.6) depends explicitly on j. This possibility is

---

[1] We do not consider here the networks which include elements with a memory. If the memory is not very long it can be taken into account by enlarging the set of state variables.

realized, for instance, in neural networks. However, there are also many situations when the network is completely uniform, which implies that the properties of all elements and of all connections are identical. In this case there is a single universal mapping function G for all elements, i. e. the rules of transition are the same for any element.

The above definition does not require the network to be regular. It can have a very complicated topology due to the links that connect any given element with many others, which may be very distant. There is however an important class of uniform networks, which are called *cellular automata*, where the elements occupy the sites of a regular lattice (square or hexagonal, simple cubic, etc.). In a cellular automaton any element is connected only with its nearest neighbors in the lattice.

The total number of all possible cellular automata is surprisingly large, even if we consider only the elements with relatively few discrete states.

Suppose that any single element can be found in one of K different states and that every element has exactly r neighbors. What is the total number of different transition rules for such a network?

According to (1.1.6) the next state of a given element is determined by $r + 1$ integers each taking K different values, i. e. by the current state of a given element and the states of its r neighbors. Hence we have K^{r+1} possible combinations. When a transition rule is specified, we should put into correspondence with every such combination a certain state of the central element at the next time step. Since this state can be chosen from K possibilities, the total number of all possible transition rules is

$$N_{\text{cell}} = K^{K^{r+1}} . \tag{1.1.7}$$

This number is extremely large. For instance, if we consider a square lattice with only 4 neighbors and assume that the elements which occupy its sites have only 2 possible states, we find from (1.1.7) that $N_{\text{cell}} = 2^{32}$.

The dynamics of cellular automata with a finite number of states is usually irreversible. This follows from the fact that, to define a transition rule, we should map K^{r+1} different states of the neighborhood into only K possible next states of the central element. Obviously, such a mapping will always be highly degenerate, with different combinations corresponding to the same outcome. Then it is often impossible to reconstruct the initial state of the entire network uniquely from its final activity pattern.

Among various cellular automata one can find networks both with very simple and with very sophisticated dynamics, so much so that they even look like some kind of random process. Extensive simulations of different cellular automata were performed by *Wolfram* [1.10, 11] who also suggested classification scheme based on the dynamical properties of such networks.

A particularly interesting example of a cellular automaton is provided by the *Game of Life* which was introduced by *Conway* [1.12] as a mathematical entertainment. The rules of the game are very simple. Each element can be in one of two states, rest or excitation. An element goes from the state of rest into the excited state if it has three excited neighbors (all elements occupy the sites of a square lattice in which each site has four neighbors). The state of excitation is prolonged for one more time step if an element has two or three excited elements in its neighborhood. The *Game of Life* is characterized by an especially complex dynamics. It generates sophisticated sequences of patterns which are extremely sensitive to the initial conditions.

One can equally well consider the automata networks where the states of individual elements are described by continuous variables or by sets of such variables.

In some situations, interactions between the elements are relatively weak and the momentary states of the neighboring elements do not differ significantly. Then an approximation can be used in which we specify the momentary state of the entire system by a continuous distribution $a(r, t)$ of its activity as a function of the coordinates r. If furthermore such a distribution changes very little at each individual time step, then its temporal evolution can be modelled by some partial differential equation. In this way we come to the concept of an *active medium*.

Since we assume that the interactions between elements are local and that the spatial variation of a is slow, the right side of an evolution equation can depend only on the local value of a and on its lowest spatial derivatives, which yields

$$\dot{a} = w(a, \text{grad}\, a, \Delta a) \, . \tag{1.1.8}$$

Moreover, we can decompose the right side in terms of grad a and Δa, keeping only the first terms:

$$\dot{a} = g(a) + A\, \text{grad}\, a + B(\text{grad}\, a)^2 + D\Delta a \, . \tag{1.1.9}$$

One can easily check that both last terms have the same order in the inverse characteristic length that specifies the scale of spatial variation. Generally, coefficients A, B, and D represent certain functions of a.

When the medium is isotropic, evolution equations should be invariant under rotations of the coordinate system and under reflections. Since the gradient of a changes its sign under the reflection $r \rightarrow -r$, it cannot enter linearly into the evolution equation of isotropic media, i.e. the second term on the righthand side of (1.1.9) should be absent.[2]

It occurs rather rarely that an active medium can be well described by a single coordinate-dependent variable. Usually, we have to consider models where the momentary state of any small element is specified by a set of several components $a_i(r, t), i = 1, 2, \ldots M$. The above arguments are easily generalized to this case. Instead of (1.1.9), for an isotropic multicomponent active medium we find

$$\dot{a}_i = g_i(\{a_j\}) + \sum_j B_{ij}(\text{grad}\, a_j)^2 + \sum_j D_{ij}\Delta a_j \, . \tag{1.1.10}$$

General equations (1.1.10) describe a great number of different active media. However, current theoretical investigations are centered on a few simple special classes of such models.

---

[2] A special situation arises when the elements of the medium can move. Then the right side of (1.1.9) can include the linear convective term $v \text{grad}\, a$, where $v(r, t)$ is the velocity field.

When the second term involving gradients is absent and $D_{ij} = D_i \delta_{ij}$, (1.1.10) takes the form

$$\dot{a}_i = g_i(\{a_j\}) + D_i \Delta a_i . \tag{1.1.11}$$

These equations describe the so-called *reaction-diffusion models*. This name is explained by the fact that equations in this class may also be used to describe the macroscopic kinetics of chemical reactions.

Suppose we have M different substances with local concentrations $a_i(\mathbf{r}, t), i = 1, 2, \ldots, M$. Inside each small volume element of the medium certain chemical reactions take place which result in changes in the local concentrations. The reaction rates are then described by a set of nonlinear functions $g_i(a_i, \ldots, a_M)$ in the right side of (1.1.11). Furthermore, the reacting molecules can diffuse to nearby volume elements and this process is described by the last term in (1.1.11), where D_i then represents the diffusion constants of the different reacting substances.

Although the "chemical" interpretation of reaction-diffusion models is very convenient, one should always remember that such models can naturally arise in many other applications which are very far from chemical kinetics. Actually, we can consider variables a_i just as a set of order parameters for a certain active medium. In some cases they might be interpreted as representing certain chemical concentrations, but these variables might also specify the local values of temperature, electric potential, population densities of some microorganisms, etc.

Active media can be further classified in terms of the behavior of an isolated element which is specified by equations (1.1.11) without the diffusion term. We can distinguish *bistable*, *excitable*, and *oscillatory* media.

In bistable media every small isolated element has two stationary states that are stable under small perturbations. Large enough perturbations can, however, trigger transitions between these two steady states. In a continuous medium composed of such elements one can observe propagation of trigger waves which represent the waves of transition from one stationary uniform state into another. These phenomena are discussed in Chap. 2 for the simplest one-component models of bistable media.

An excitable element has a single stationary state that is stable under small perturbations. However, such an element differs from a passive one in its reaction to sufficiently intensive perturbations. If the perturbation exceeds a certain threshold, this element produces a strong burst of activity. It performs a definite sequence of transitions and later returns to the initial state of rest.

Suppose we have a chain that consists of excitable elements. What might be the interaction between such elements? It is most natural to assume that an element which goes from the state of rest into the active form remains unreceptive to external influences until it completes the prescribed sequence of internal transitions. Then it is sufficient to consider only the case when we have two neighboring elements, one in the active form and the other in a state of rest. We can imagine two kinds of interaction. First, it might happen that any element in the active form can force out the neighboring element from the state of rest. Then even a single external perturbation would be sufficient to produce a persistent wave-like activity in this chain. However, this situation is uncommon. Usually, only the elements in the first stages of the activity burst are able to act on their neighbors. Such special interaction results in the propagation of the solitary *excitation pulse* along the chain, with the burst of activity relayed from one element to another. When an excitable medium is two- or three-dimensional such pulses transform into excitation waves which can produce very complicated patterns. Various properties of excitable media are analyzed in Chap. 3.

An oscillatory medium consists of elements that perform stable limit-cycle oscillations. If we take a chain of such elements and establish a constant shift in the initial phase of oscillations between any two subsequent elements, this would result in a time-dependent activity pattern which looks like a propagating wave. A well-known example of this phenomenon is given by the "waves" in a festoon of electric bulbs: each of them is periodically switched on for some time but the moments of switching are shifted for the neighboring bulbs. When interactions between the oscillatory elements are taken into account, they can modify local oscillation frequencies and change the phases of oscillations. Furthermore, very complicated wave patterns can be also observed. These effects are discussed in Chap. 4.

In Chap. 5 we consider the properties of stable stationary *dissipative patterns* that are typically found in active media where short-range activation is combined with long-range negative feedback.

In contrast to active media, which can be viewed as continuous approximations of lattice networks where every element is connected with only a small number of its nearest neighbors, *neural networks*, studied in Chap. 6, have very high connectivity, with direct connections spreading even to very distant elements. The network created by a system of such multiple connections has a complicated topology. Moreover, the properties of individual connections can also be different, so that the network is highly nonuniform. Some of the connections are activatory, others are inhibitory.

An analytical treatment of an arbitrary neural network, where both the topology and the properties of connections are chosen at random, is an excessively difficult task. Much more amenable to theoretical investigation are the models of neural networks which are specifically designed to realize various functions of analog distributed information processing. The importance of such specialized networks for the operation of the brain and their prospective applications in computing machines explain the persistent interest in neural network models.

In theoretical studies, each individual element (each *neuron*) is usually modeled by a simple two-state discrete automaton. Therefore, from a general point of view, neural networks are automata networks with a complicated topology of long-range activatory and inhibitory connections.

Reproductive networks, discussed in Chap. 7, are also characterized in some cases by a complicated pattern of connections. The elements of such networks can represent populations of certain species or outputs of given productive agents, specified by continuous variables. These models of distributed active systems are typical for the situations found in evolutionary biology, mathematical ecology and the social sciences.

1.2 Engineering and Control of Active Systems

The incentive to study distributed active systems comes not only from a desire to explain the natural forms of cooperative behavior. It now seems evident that the mathematical theory of such systems might have very promising applications in technology and information processing.

In effect, the fundamental principles of today's technology do not differ greatly from the organization of manual labour. As we know, the process of individual manual production consists in a sequence of operations with tools, each tool being a passive object independent of the others. A human formulates a logical program of actions with these tools and executes them in the required order. Execution of a production task is preceded by a rational analysis which results in decomposition of the task into a sequence of elementary operations.

The presence of a central organizing agent is crucial, as well, for any modern industrial production line. There the operations of individual workers are performed sequentially in accordance with a certain externally imposed plan, and are themselves independent. Moreover, even in the most advanced fully automatic machines we can again see that there is an external program of sequential operations which is recorded in their design.

This kind of production process is different in principle from the synergetic forms of behavior that are natural for living beings. The usual behavior of a living system is based on cooperative interactions between many units which result in complex self-organization. Such a mode of operation does not presume the existence of an external agent who would have a complete plan of actions and would supply orders to individual units. Instead, the behavior of a living system follows from interactions between its units.

Hence, a synergetic "program of operation" consists simply in the laws of interaction that govern the dynamics. To change the behavior of a system, we have to modify these laws.

Investigations of mathematical models of distributed active systems show that very simple active units are able to produce, through cooperative interactions, ordered behavior of high complexity. Even more complex behavior can be observed in a hierarchy of self-organized subsystems.

It seems promising to implement the principle of self-organization in industrial technological devices. Effectively, this would mean the introduction of artificial "living systems" into the technological chain. In a distant perspective, one can expect (partial) replacement of the traditional "grey" technology by a synergetic "green" technology, as this was discussed recently by *Dyson* [1.13]. Note that the term "green" does not imply a biological origin for new devices, but rather characterizes the manner of their operation, largely autonomous and self-organized.

Obviously, these future applications are possible only if a science of *engineering* distributed active systems, which allows the construction of such systems and their hierarchies with the desired self-organization properties, is developed.

A closely related problem is how to *control* complex active systems. For a long time it was believed that an ideal means of control is to supply detailed instructions to each of the elements. Such strict instructions should be issued by some central planning agency, presumably proceeding from the complete data obtained from all elements and the known laws of their operation. However, the vast (and sometimes rather sad) experience shows that this scheme of control is not efficient for large interactive systems and inevitably leads to instabilities and chaotic behavior.

The mathematical aspects of this problem will be discussed in the second volume. In the meantime we simply note that, even when a system has some inherent capabilities for self-organization, these are destroyed by intensive controlling interference. Small perturbations develop into large-scale random variations and this cannot be prevented by a central agency. It never receives the complete correct data and its instructions always arrive with certain delays. Hence, trying to damp the fluctuations, it effectively enhances them if the system is sufficiently complicated.

Another approach is realized in living systems. As already noted, these systems are made up of hierarchies of simple self-organized subsystems. Here the control coming from the higher levels never directly interferes with the basic dynamics of individual elements; instead it consists in weak influences that slightly modify the laws of interaction between the elements of the lower level and change the properties of patterns which are formed autonomously by these elements. This leaves the task of maintaining stability in response to external perturbations to the subsystem itself.

In biological systems such a synergetic control mechanism is automatically worked out in the process of evolution. To realize this mechanism for artificial active systems, one should first discover the laws which determine the autonomous formation of patterns in such systems. Note that this is different in principle from the kind of knowledge required for centralized control by an external organization, where only the individual dynamics of isolated elements is significant. Hence, we come to the same sort of problems as in the engineering of distributed active systems.

A remarkably similar situation is found in the field of *information processing*.

When a man performs some computation he acts in effectively the same manner as in the process of manual labour. Numbers are taken from the memory (either internal or external) and then returned to it after completion of each individual arithmetical operation. The sequence of such operations is determined by a mental plan, or a *program*, which had been formulated before the computation started.

The abacus and other simple computing devices are, in effect, tools that allow the mechanization of individual arithmetic operations, but that still require the active participation of a human. A great step forward was made with the introduction of computing machines, which we are so proud of. These machines do not require direct human participation. Instead, they automatically execute a sequence of mathematical operations recorded in the program which was fed into it by a human. Actually, modern computers act in the same manner as industrial fully automatized production machines. Both presume the presence of some external agent who has already logically decomposed the problem into a sequence of elementary operations.

Obviously, information processing also includes many other tasks, in addition to numeric calculations. However, traditional computers can only treat such problems when they are effectively reduced to sequential computational models.

Again, this mode of information processing is absolutely strange for living beings. Nobody would seriously claim that, when a cat jumps to catch a mouse, its brain solves by computation a complicated set of differential equations that describe the trajectory of its motion. It is equally absurd to assume that a goalkeeper computes mentally the trajectory of his jump to intercept the flying ball. The faculty of rational reasoning has emerged in humans by a gradual evolution of the older and more fundamental mechanisms of information processing, present in more primitive animals. Moreover, it is used, in its pure form, only in a few rather special situations.

An alternative to computation is provided by the *analog* processing of information. This is based on the idea of mapping the problem onto the dynamics of some artificially constructed system, so that the answer can be obtained by following the evolution of this system.

In the middle of our century, when universal computers were not yet available, flight trajectories of missiles were often predicted by using specialized analog machines. These electronic devices included circuits in which the dynamics of the electric current obeyed effectively the same differential equations as those which describe the flight of a missile. Hence, it was possible to imitate any particular flight by adjusting the initial conditions and then observing the electric current dynamics in the circuit.

This example illustrates one possible application of analog machines. We see that they can represent direct physical models, which employ other elements but still have the same laws of evolution as the system to be modeled. By using distributed active systems as analog devices, one can directly imitate in this way the processes which occur in various media.

However, even if we assume that the fundamental mode of activity of the brain is of an analog nature, it cannot be reduced to direct imitation of the surrounding world.

Obviously, the brain cannot model all the outside processes that are reflected in the vast information flow coming from our senses. Instead it should use some internal representation of the surrounding world which includes only those properties which are essential for the life of a human or an animal. Firstly, this implies some primary processing of the information flow which allows the brain to classify different states of the environment. Such classification can be successful only if it is based on definite invariants in the surrounding world. The search for such *invariants* (or regularities) in the surrounding world is one of the most important functions of the brain.

For instance, all letters A, independent of the script, should be recognized as the same object; all triangles with various orientations and aspect ratios should be considered as objects different from squares or circles. We can recognize persons, i.e. we are able to determine that we see the same man, independent of his age, clothes, face expression, etc. We are also able to identify temporal sequences: it is not usually difficult for us to catch the melody – independent of whether it is played by a music instrument, the human voice, a choir of voices, or an entire symphonic orchestra.

The classification ability of the brain is called the property of *associative memory*. Almost certainly, this property is realized in the brain in an analog manner. It can also be easily implemented in artificially constructed active systems.

Suppose that an active system has several stable attractive stationary patterns. Each of these patterns has its own attraction basin. If the initial state of the system falls into some particular basin, its evolution will result in approaching the corresponding pattern.

Suppose further that any initial activity pattern of the system corresponds to some possible picture which we want to classify and that each of the attractive patterns defines a certain prototype picture. Since any initial pattern should belong to one of the attraction basins, it will evolve in time into the corresponding prototype. In this way an analog associative memory can operate.

This simple idea, used already by *Haken* [1.14], establishes a connection between the property of associative memory and pattern formation in distributed active systems. Recently, much attention has been paid to the models of associative memory which utilize a particular kind of such systems, namely artificial neural networks. We discuss these models in detail in Chap. 6. In the same chapter the reader will find a description of simple neural networks that are able to store and retrieve complex temporal sequence.

Nevertheless, neural networks are not the only kind of active system endowed with these properties. Therefore, a device with an associative memory can equally well be realized by using distributed active systems of some other origin. In Sect. 7.5 we briefly discuss models of associative memory which are based on reproductive systems.

It should be emphasized that the actual operation of the brain, involving classification of patterns, is never reduced to a single application of the analog classification procedure, as this was outlined above. It always consists of several stages and has a clear hierarchical character.

Consider, for instance, the sequence of steps which can be used to distinguish a triangle from all other polygons. The first step consists in identification of straight line segments. Then it is necessary to identify vertices, i.e. the points where two lines meet. In principle, we can realize both these steps in an analog manner, by using distributed active systems with local associative memory. At the next stage it is necessary to construct an intermediate, compressed representation of the picture, i.e. its graph. Such a graph indicates only the presence of links between the vertices, independent of the length or the orientation of the lines which realize the connections. Now, since each graph corresponds to an entire class (e.g. to all possible triangles), it is only necessary to recognize the graph in question to give a final answer to the problem.

We see that even the solution of such a simple classification problem requires its translation into some internal "language". The fundamental elements of such a language represent already certain abstract concepts (in the above example they are the vertices and the links realized by straight line segments). These elements form patterns which are called *semantic nets*. Each such net (e.g. the graph of a triangle) can in its turn be considered as a new concept, and in this guise it can enter into a semantic net at the next hierarchical level.

It seems highly plausible that the principal mode of any advanced brain functioning consists in different operations with semantic nets that include their recognition, generation, transmission, transformation, and comparison. These operations with semantic nets should be present in all animals, varying in their complexity. Hence, they have to be realized in an analog manner, as dynamic processes in the neural network of the brain. In other words, the brain can be viewed as a special distributed active system that provides a living space for semantic nets which evolve, interact and compete with one another (see [1.15]). The theoretical problems of analog graph processing by distributed active systems are not sufficiently well investigated. At present, this field of research in the mathematical theory of active systems seems to be very promising, as seen from the perspective of the construction of large-scale analog machines.

We know that the brain, which is a huge distributed active system, can also very effectively control the behavior of other active systems. This is most clearly manifested in its control of motor activity, i.e. of muscle contractions. No robot, controlled by even the most advanced modern computer, can reproduce the very subtly coordinated motions of a man.

The total number of independent degrees of freedom in typical robots is only about 10, whereas the number of motor units involved in the formation of motion in humans and higher animals can reach 10^5. Evidently, this large system can be controlled only if it is hierarchically structured. Moreover, the same should be true for the neural system which forms the control signals. The initial "idea" of motion should develop itself into a hierarchy of self-organized activity patterns that specify all the details of the complex motion.

There is still another aspect in the problem of analog information processing by distributed active systems. Since the program of operation of analog machines is recorded in the laws of interactions between different elements, i.e. built into its "hardware", such machines are not universal. Although there are several very important tasks that might justify the construction of certain narrowly specialized machines, generally this nonuniversality represents a serious deficiency.

Such a deficiency can, however, be effectively overcome if we endow such a machine with the ability to *learn*. Suppose that initially the machine has only one built-in program, i.e. the learning program. At the initial stage the machine would construct its internal structure by establishing new connections and-or breaking old connections between the elements, by modifying the properties of individual elements, etc. When learning is finished, it would have acquired the ability to solve a special kind of problem. Later the learning process can be repeated, allowing use of the same machine for other tasks.

2. Bistable Media

Bistable media consist of elements that have two steady states which are stable under sufficiently small perturbations. Stronger perturbations can cause transitions between these states. The fundamental form of a pattern in bistable media is a trigger wave, which represents a propagating front of transition from one stationary state into the other. The propagation velocity of a flat front is uniquely determined by the properties of the bistable medium. To initiate a spreading wave of transition from a homogeneous steady state, one should create a local perturbation which exceeds a critical nucleus for the bistable medium.

2.1 One-Component Models of Bistable Media

Bistable behavior is found in different media, the elements of which can be rather complex and require description by sets of many variables. However, all principal properties of patterns in bistable media are present already in the simplest one-component models, where each individual element is described by a single equation

$$\dot{u} = f(u) \tag{2.1.1}$$

with the nonlinear function $f(u)$ shown in Fig. 2.1.

Note that any root u^* of $f(u) = 0$ corresponds to a possible stationary state of the element. This stationary state is stable if all small deviations δu from u^* are damped. Since small deviations obey a linearized equation

$$\delta\dot{u} = f'(u^*)\,\delta u \ , \tag{2.1.2}$$

a stationary state u^* is stable only if the derivative $f'(u^*)$ is negative. For the function $f(u)$ shown in Fig. 2.1, we have two stable states u_1 and u_3 and one unstable state u_2.

As the first example, consider an active element which represents a cell with fuel. For simplicity we assume that the amount of fuel does not change in time (either because new portions of fuel are permanently supplied, or because its store is so large that we can neglect the decline due to consumption). Combustion of fuel is accompanied by heat production at a rate q. The dependence of q on temperature θ should be step-like (see Fig. 2.2): at low temperatures combustion is absent and no heat is produced.

---

[1] Some additional effects observed in two-component bistable media will be discussed in Chap. 5.

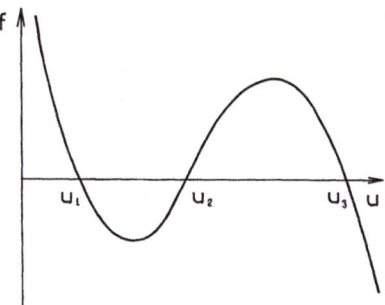

Fig. 2.1. The function $f(u)$

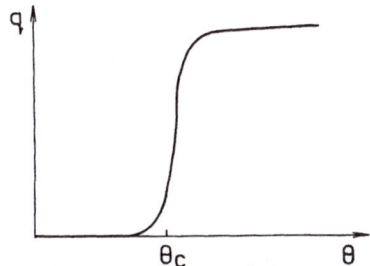

Fig. 2.2. Typical dependence of heat production rate q on temperature θ

If an element is isolated, all combustion heat is used for heating. Therefore, for a small time interval Δt at the increment $\Delta\theta$ of temperature is given by

$$q(\theta)\,\Delta t = C\Delta\theta\,, \tag{2.1.3}$$

where C is the heat capacity. The continuous evolution of temperature is then described by the equation

$$\dot\theta = (1/C)\,q(\theta)\,. \tag{2.1.4}$$

Since $q(\theta)$ is nonnegative, temperature increses indefinitely.

In a realistic situation, however, there is always a certain heat exchange between an element and its environment that sets a limit to increasing temperature. If the temperature of the environment is θ_1 and the heat transfer coefficient is γ, the modified equation taking into account the effects of heat exchange reads

$$\dot\theta = (1/C)q(\theta) - \gamma(\theta - \theta_1)\,. \tag{2.1.5}$$

Introducing $f(\theta) = (1/C)q(\theta) - \gamma(\theta - \theta_1)$, we can write this in the form (2.1.1)

Stable stationary states of such a system have a very simple interpretation. In the case shown in Fig. 2.3a, the cold state θ_1 corresponds to the absence of fire. In the stationary hot state θ_3 all the heat produced is removed at the same rate from the cell. Both states are stable with respect to *small* perturbations. However, larger perturbations can lead to transitions between states; it is possible to stoke the fire up or to extinguish it.

Under a different relationship between the parameters, the same system might possess a single stable stationary state. If the heat production is too low or the produced heat is removed too quickly (see Fig. 2.2b), there is no steady combustion

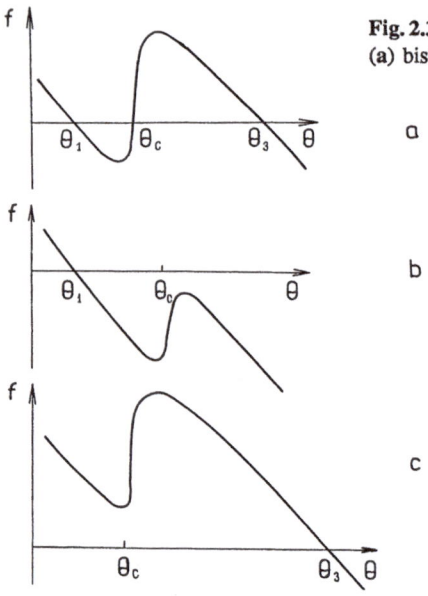

Fig. 2.3a-c. Three possible situations for a cell which contains fuel:
(a) bistability, (b) ignition is impossible, (c) self-ignition

regime and the only steady state is θ_1. On the other hand, if the temperature θ_1 of
the environment exceeds the critical temperature θ_c of ignition (Fig. 2.2c), the cell
is self-igniting. Then it has only the hot steady state θ_3.

The combustion process in a continuous medium filled by fuel is described by
the equation

$$\frac{\partial \theta}{\partial t} = f(\theta) + \chi \frac{\partial^2 \theta}{\partial x^2} , \tag{2.1.6}$$

where χ is a heat conductivity coefficient. When each of its isolated small cells has
two steady states, this active medium is bistable.

Another typical example of a one-component bistable medium is found in a situation when we have a threshold production of a certain substance, counterbalanced
by a process which tends to decrease its concentration.

Consider a hypothetical trimolecular chemical reaction with an autocatalytic
stage, which was proposed by *Schlögl* [2.1], i. e.

$$A + 2X \underset{k_2}{\overset{k_1}{\rightleftharpoons}} 3X , \quad X \underset{k_4}{\overset{k_3}{\rightleftharpoons}} B . \tag{2.1.7}$$

If we suppose that concentrations of the reagents A and B are fixed, this system
will be described by a single kinetic equation for the concentration of the reagent
X, i. e.

$$\dot{X} = k_1 A X^2 - k_2 X^3 - k_3 X + k_4 B . \tag{2.1.8}$$

The nonlinear function $f(X)$ defined as the right hand side of this equation is a
cubic polynomial. In a certain interval of the parameters k_1, k_2, k_3, k_4, A, and B,
this function has the form shown in Fig. 2.1, and therefore such an irreversible

chemical reaction can possess two stable steady regimes that are characterized by different values X_1 and X_2 of the concentration X.

The kinetic equation (2.1.8) is written under the assumption of ideal diffusion mixing of the reagents within a given cell. If we take a chain consisting of such cells, there will be an exchange of molecules between the neighboring cells due to diffusion. In the continuum approximation (i. e. for the continuous medium) the spatial distribution of X evolves in time according to

$$\frac{\partial X}{\partial t} = f(X) + D\frac{\partial^2 X}{\partial t^2} \, , \qquad\qquad (2.1.9)$$

where D is the diffusion constant.

A closely related model describes the time evolution of an ecological system. Suppose that some insects with local population density n reproduce in the medium. Let us assume that their reproduction is sexual, i. e. that the birth rate is proportional[2] to the probability of an encounter between a male and a female, namely to n^2. Taking into account also the possibility of death for insects and their diffusional wandering in the medium, we obtain an equation

$$\frac{\partial n}{\partial t} = -\gamma n + \alpha m(n)n^2 + D\Delta n \, . \qquad\qquad (2.1.10)$$

The birth rate is proportional to the amount of food m which depends on the population density of the insects. Generally, the food dynamics should be described by a separate evolution equation. We limit our discussion to a special case where food rapidly adjusts to the population density of the insects and follows it adiabatically. Under this condition, the amount of food is a decreasing function of n. To prevent an infinite growth of the population, the decline in m should be sufficiently steep, e. g.

$$m = m_0 \exp(-n/n_0) \, . \qquad\qquad (2.1.11)$$

Such an ecological system represents a bistable medium with a nonlinear function $f(n)$ given by

$$f = -\gamma n + \alpha m_0 \exp(-n/n_0)n^2 \, . \qquad\qquad (2.1.12)$$

This medium has two stable uniform states. The first of them is $n = 0$. This state is stable because the probability of meeting of two insects, one male and one female (and therefore the birth rate) is vanishingly small at low population densities. Saturation of growth due to food exhaustion guarantees the stability of the second stationary state.

Thus, we see that bistable one-component media described by the equation

$$\frac{\partial u}{\partial t} = f(u) + D\Delta u \, , \qquad\qquad (2.1.13)$$

---

[2] This model is very simplified. For instance, we assume that local population densities of males and females are always the same.

are found in diverse applications. Further particular examples are given, for instance, in the monograph [2.2] and in the reviews [2.3, 4].

2.2 Trigger Waves

The basic kind of pattern in bistable media is a *trigger wave*; its propagation triggers the transition from one stable stationary state of the medium to the other. Such waves were first observed in 1906 in a chemical reacting medium by *Luther* [2.5], who also made theoretical estimates of their propagation velocity. Their detailed mathematical description was given in 1937 by *Fischer* [2.6], who studied a problem in population genetics, and one year later by *Zeldovich* and *Frank-Kamenetskii* [2.7] in the theory of flame propagation.

A trigger wave which realizes the transition from a state u_3 to a state u_1 and moves at a velocity c is given by a special solution

$$u = u(\xi), \quad \xi = x - ct \tag{2.2.1}$$

that satisfies the boundary conditions

$$\begin{aligned} u &\to u_3, \quad \xi \to -\infty, \\ u &\to u_1, \quad \xi \to +\infty. \end{aligned} \tag{2.2.2}$$

After substitution of (2.2.1) into (2.1.13) we obtain the equation

$$-cu' = f(u) + Du'', \tag{2.2.3}$$

where the primes denote derivatives with respect to ξ.

Although further analysis can be performed in abstract terms, it is more instructive to use a simple mechanical analogy. Let us introduce a function

$$U(u) = \int_0^u f(u)du \tag{2.2.4}$$

and rewrite (2.2.3) in the form

$$Du'' = -\frac{\partial U}{\partial u} - cu'. \tag{2.2.5}$$

If we interpret u as a coordinate of some particle and ξ as time, (2.2.5) reduces to the equation of motion of this particle of mass D in a potential $U(u)$ in the presence of viscous friction proportional to the velocity u' of the particle. Hence, in this analogy c plays the role of a viscous friction coefficient.

The potential $U(u)$ has two maxima at u_1 and u_3 and a local minimum at the intermediate point u_2 (Fig. 2.4). Let us assume that $U(u_3) > U(u_1)$. According to (2.2.4), this condition holds if $A > 0$, where

$$A = \int_{u_1}^{u_3} f(u)du. \tag{2.2.6}$$

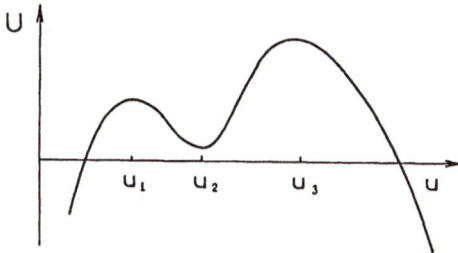

Fig. 2.4. Dependence of potential U on coordinate u

If viscous friction is absent (i. e. if $c = 0$), (2.2.5) conserves the energy

$$E = U(u) + \frac{1}{2} D \left(\frac{du}{d\xi} \right)^2 . \tag{2.2.7}$$

In this case, after being released with a vanishingly small velocity from the point u_3 where the potential U has its higher maximum (Fig. 2.4), the particle arrives after a finite time ξ at the lower maximum of U at u_1. Since energy is conserved, the particle passes u_1 with finite velocity and continues to move farther in the direction of negative values of u. Hence, conditions (2.2.2) cannot be satisfied if $c = 0$.

On the other hand, if the viscous friction coefficient c is too large, the particle rapidly loses its energy and cannot climb the maximum at u_1. Consequently, as $\xi \to \infty$ it will stay at the point u_2 where U has a minimum. This also violates (2.2.2).

For a given potential $U(u)$, there is only one value c_0 of the viscous friction coefficient c for which the dissipative loss of energy is exactly equal to the difference of the potentials in the points u_1 and u_3, i. e. to $\Delta E = U(u_3) - U(u_1)$. Then, after being released at u_3, the particle passes the minimum at u_2 and for an infinitely long time climbs up to the maximum at u_1, arriving there with zero velocity. This special value represents the propagation speed c_0 of a trigger wave.

Thus, both the propagation speed of a trigger wave, and its profile are uniquely determined by the properties of the medium. All trigger waves in a given medium have the same profile and velocity, independent of the conditions that lead to their creation.

Note that c_0 diminishes for smaller values of ΔE and vanishes when $U(u_3) = U(u_1)$. If $U(u_3) < U(u_1)$, a trigger wave moves in the opposite direction, i. e. we have $u \to u_1$ for $\xi \to -\infty$ and $u \to u_3$ for $\xi \to \infty$. Therefore, when A is negative, propagation of a trigger wave results in a transition from u_3 to u_1.

The same conclusions can be derived in a more formal way, without resorting to mechanical analogies. Figure 2.5 shows phase trajectories of equation (2.2.5) on the plane (u, u') when $A < 0$. For $c > c_0$, a trajectory starting from the stationary point $(u_1, 0)$ is absorbed by another stationary point $(u_2, 0)$, whereas for $c < c_0$ it goes to infinity. According to a standard classification, $(u_3, 0)$ and $(u_1, 0)$ are the saddle points of (2.2.5), and the trajectories passing through these points are their separatrixes. Generally, separatrixes of the two saddle points are different. Only at $c = c_0$ is there a single separatrix trajectory that leaves one saddle point and reaches the other. This special trajectory corresponds to a trigger wave.

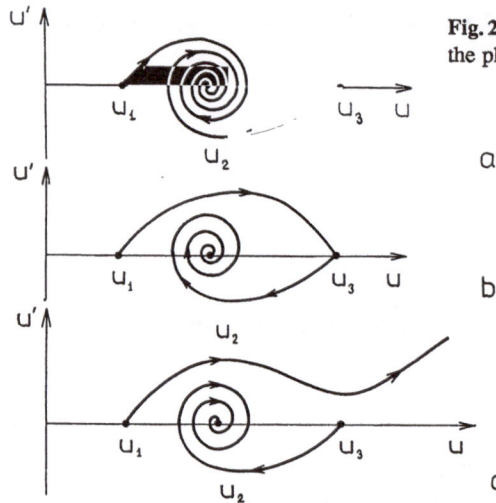

Fig. 2.5a-c. Some phase trajectories of equation (2.2.5) on the plane (u, u') for (a) $c > c_0$, (b) $c = c_0$, and (c) $c < c_0$

No general analytical methods for calculation of the propagation velocity for an arbitrary function $f(u)$ are available. Nevertheless, in an important special case when the function $f(u)$ represents a cubic polynomial,

$$f = -\kappa(u - u_1)(u - u_2)(u - u_3), \quad u_1 < u_2 < u_3 , \tag{2.2.8}$$

there is an exact analytical solution of this problem. By substituting the expression $du/d\xi = B_0(u - u_1)(u - u_3)$ into (2.2.3), one can find a solution satisfying the boundary conditions (2.2.2) only at $B_0 = (\kappa/2D)^{1/2}$. Then, the propagation speed of a trigger wave is

$$C_0 = \tfrac{1}{2}\sqrt{\kappa D}(u_1 + u_3 - 2u_0) . \tag{2.2.9}$$

Analytical solutions can also be found in models with a piecewise-linear approximation of an S-shaped function $f(u)$. Suppose, for example, that a medium is described by (2.1.13) with

$$f(u) = -u + (u_3 - u_1)H(u - u_2) + u_1 , \tag{2.2.10}$$

where $H(z) = 1$ for $z \geq 0$ and $H(z) = 0$ for $z < 0$. This medium has two stable states u_1 and u_3 and an unstable state u_2.

If $u_1 + u_3 > 2u_2$, a trigger wave realizes a transition from u_1 to u_3. Since $f(u)$ is piecewise-linear, a special solution for a trigger wave propagating at a velocity c can be sought in the form

$$u = C_1 \exp[-(\lambda_1/c)\xi] + u_1, \; \xi > 0 ,$$
$$u = C_3 \exp[-(\lambda_3/c)\xi] + u_3, \; \xi < 0 . \tag{2.2.11}$$

Here λ_1 and λ_3 are given by the roots of a characteristic equation obtained by substitution of $u \sim \exp[-(\lambda/c)\xi]$ into (2.2.3),

$$\lambda_{1,3} = (c^2/2)\left[1 \pm \sqrt{1 + 4/c^2}\right] . \tag{2.2.12}$$

Coefficients C_1 and C_3 can be found from the condition that $u = u_2$ at $\xi = 0$. By matching the left and the right derivatives of u at $\xi = 0$, we obtain an equation for the propagation velocity c that has a solution

$$c_0 = \frac{u_1 + u_3 - 2u_2}{\sqrt{(u_2 - u_1)(u_3 - u_2)}} . \tag{2.2.13}$$

When $u_1 + u_3 < 2u_2$, a trigger wave realizes a transition from u_3 to u_1. The propagation velocity is again given by (2.2.13).

There are some approximate methods that can be used to determine the properties of trigger waves in the general case.

Let us first consider *slow* trigger waves. As shown above, the propagation speed of a trigger wave vanishes (and a trigger wave becomes a static interface) when $A = 0$. It can be expected that the propagation speed c will be small for small values of A.

If we multiply both sides of (2.2.3) by $du/d\xi$, integrate over ξ from $-\infty$ to ∞, and use conditions (2.2.2), we find

$$c \int_{-\infty}^{\infty} \left(\frac{du}{d\xi} \right)^2 d\xi = \int_{u_1}^{u_3} f(u) du . \tag{2.2.14}$$

This equation can be used to estimate the propagation speed c.

When c is small, the profile $u(\xi)$ of a propagating wave is very close to the profile $u^{(0)}(\xi)$ of a static interface found for $A = 0$. Replacing $u(\xi)$ in (2.2.14) by $u^{(0)}(\xi)$ and taking into account the definition (2.2.2) of A, we find approximately

$$c = A \left\{ \int_{-\infty}^{\infty} \left(\frac{du^{(0)}(\xi)}{d\xi} \right)^2 d\xi \right\}^{-1} . \tag{2.2.15}$$

For $A = 0$, the profile $u^{(0)}(\xi)$ of a static interface is given by the equation

$$Du'' = -\frac{\partial U}{\partial u} . \tag{2.2.16}$$

This has an analytic solution

$$\xi = \int_{u_1}^{u} \left\{ \left(\frac{2}{D} \right) [U(u_1) - U(u)] \right\}^{-1/2} du \tag{2.2.17}$$

that expresses ξ as a function of u for a static interface.

Equations (2.2.15) and (2.2.17) determine the propagation speed of slow trigger waves. Note that this speed is proportional to A.

Another approximate method was proposed in 1938 by *Zeldovich* and *Frank-Kamenetskii* [2.7] in an analysis of flame propagation.

Let us suppose that the nonlinear function $f(u)$ has the characteristic form shown in Fig. 2.6a, i.e. it reaches large positive values in a narrow interval from u_2 to u_3 and is negative, but small in magnitude, within the interval from u_1 to u_2. The

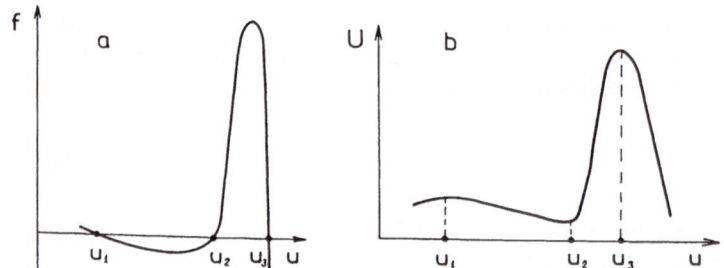

Fig. 2.6a,b. The Function $f(u)$ (a) and potential $U(u)$ (b)for fast trigger waves

corresponding form of the potential $U(u)$ in the mechanical analogy is shown in Fig. 2.6b.

Motion of a particle in such a potential can be divided into two different stages. In the first stage (for $u_2 < u < u_3$) the particle is rapidly accelerated; here the potential force $f(u)$ dominates and we can neglect the viscous friction force $-cu'$. Under this approximation the equation of motion is

$$Du'' \approx -\frac{\partial U}{\partial u} , \quad u_2 < u < u_3 , \tag{2.2.18}$$

and the total energy of the particle is conserved,

$$\tfrac{1}{2}D(u')^2 + U(u) = U(u_3) . \tag{2.2.19}$$

Hence, by the end of the first stage (i. e. at the point u_2), the velocity of the particle is

$$\frac{du}{d\xi} \approx \sqrt{\left(\frac{2}{D}\right)[U(u_3) - U(u_2)]} . \tag{2.2.20}$$

In the second stage (for $u_1 < u < u_2$) the dominant role is played by the force of viscous friction, i. e.

$$Du'' \approx -cu' . \tag{2.2.21}$$

The decelerating particle has initial velocity $(du/d\xi)_0$ given by (2.2.20) Since $u \to u_1$ in the limit $\xi \to \infty$, the deceleration path can be estimated by a direct integration of (2.2.20). Combining these two expressions, we find

$$u_2 - u_1 = \frac{D}{c}\left(\frac{du}{d\xi}\right)_0 . \tag{2.2.22}$$

Substitution of $(du/d\xi)_0$ from (2.2.20) gives

$$c = (u_2 - u_1)^{-1}\sqrt{2D[U(u_3) - U(u_2)]} . \tag{2.2.23}$$

Under our assumptions, $U(u_2)$ does not significantly differ from $U(u_1)$ and u_2 is close to u_3. Therefore, we can approximately rewrite (2.2.23) as

$$c \approx \frac{\sqrt{2DA}}{u_3 - u_1} \ .$$

(2.2.24)

This approximate expression gives the speed of propagation of fast trigger waves.

Other theoretical aspects of trigger wave propagation are discussed in the monograph by *Zeldovich* et al. [2.8].

2.3 General Properties of Patterns in One-Component Bistable Media

The analysis carried out in the previous section leaves open several important questions. Are the trigger waves stable? Are they the only possible type of propagating pattern in such media? What happens when two of these waves collide? What is the subsequent evolution of an arbitrary initial distribution?

To examine these problems, we put (2.1.13) into the form

$$\frac{\partial u}{\partial t} = -\frac{\delta F[u]}{\delta u(r, t)} \ ,$$

(2.3.1)

where the right hand side is expressed as the variational derivative of a functional[3]

$$F = \int \left\{ -U\left(u(r)\right) + \tfrac{1}{2} D(\mathrm{grad}\, u)^2 \right\} dr$$

(2.3.2)

and the function $U(u)$ is given by (2.2.4).

It follows from (2.3.1) that F never increases with time, whatever the choice of initial distribution $u(r)$. Indeed, the time derivative of F is

$$\frac{dF}{dt} = \int \frac{\delta F}{\delta u(r, t)} \frac{\partial u}{\partial t} \, dr$$

(2.3.3)

and, by using (2.3.1), we find

$$\frac{dF}{dt} = -\int \left(\frac{\delta F}{\delta u(r, t)} \right)^2 dr \ .$$

(2.3.4)

Thus we see that the time derivative of F vanishes for stationary distributions and is negative otherwise (F has the property of a Lyapunov functional).

All stable stationary distributions correspond to minima of the functional F. In the process of its evolution, F decreases until one of its minima is reached. Hence, the evolution of this system consists in approaching one of the stationary distributions.

---

[3] The variational derivative is defined as

$$\frac{\delta F[u]}{\delta u(r_1)} = \frac{\lim}{\int \delta u(r) dr \to 0} \frac{F[u + \delta u] - F[u]}{\int \delta u(r) dr} \ ,$$

where the variation $\delta u(r)$ is localized near the point $r = r_1$.

To begin our analysis, we note that the uniform distributions $u(r) = u_1$ and $u(r) = u_3$ are always stable with respect to small perturbations.

If we slightly modify such a uniform distribution by inserting a coordinate-dependent perturbation $\delta u(r)$, the change of the functional F will consist of two parts. Small variations of u increase the first term under the integral sign in (2.3.2), because uniform distributions correspond to the maxima of the potential U and U enters into (2.3.2) with a negative sign. Furthermore, in the presence of a perturbation the second term in (2.3.2) no longer vanishes and, since it is proportional to the square of the gradient of δu, it too gives a positive contribution. Since both contributions are positive, any small perturbation results in an increase of the functional F. This proves that the uniform distributions u_1 and u_3 are stable with respect to small perturbations.

Next we discuss the evolution of some strongly nonhomogeneous distributions. Suppose for a moment that a system is one-dimensional.

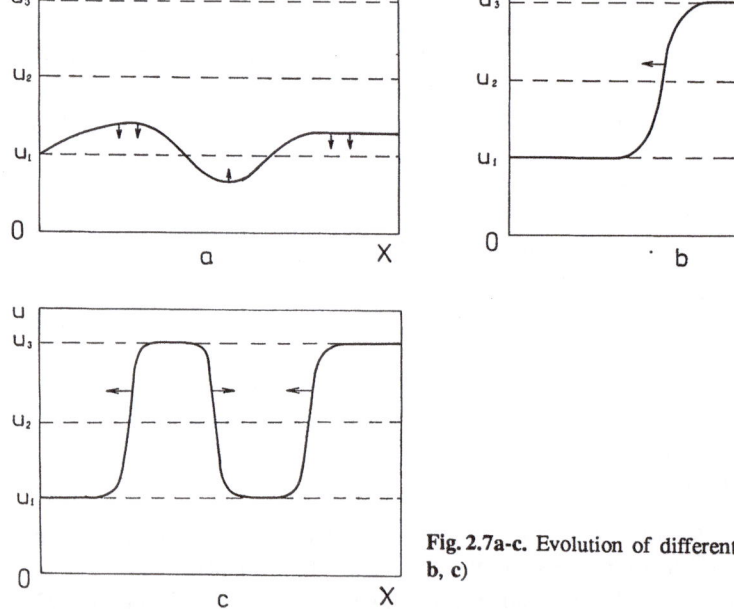

Fig. 2.7a-c. Evolution of different initial distributions (a, b, c)

Consider an initial distribution for which u never exceeds the value u_2, corresponding to a minimum of $U(u)$ (see Fig. 2.4 and Fig. 2.7a). Obviously, such a distribution cannot realize a minimum of the functional F. By moving u closer to u_1 at every point x, we decrease the first (potential) term $-U(u)$ at this point. At the same time, since this makes the distribution more uniform, it diminishes the second (gradient) term under the integral sign in (2.3.2). Therefore, such an initial distribution should relax to the uniform state $u = u_1$. By similar arguments, it can be proved that any distribution for which u is everywhere higher than u_2 is unstable and evolves to another uniform state $u = u_3$.

The initial distribution might, alternatively, have the form shown in Fig. 2.7b. Distributions of such kind are unstable because we can always decrease F by moving the interface (i. e. the boundary between the regions where u is close to u_1 or u_3), until this interface reaches a border of the medium. Thereby a stationary uniform distribution will again be achieved. It is this moving interface that represents a trigger wave.

Since any initial distribution can be divided into parts belonging to one of the two kinds discussed above (Fig. 2.7c), it always relaxes to a uniform distribution. In particular, this implies that any two colliding trigger waves should *annihilate*.

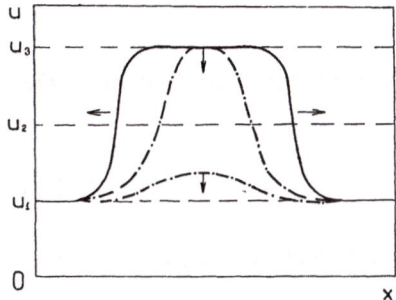

Fig. 2.8. Evolution of localized initial perturbations of a homogeneous steady state

It was noted above that both uniform states u_1 and u_3 are stable with respect to small perturbations. An absolute minimum of F is reached, however, only for one of these states, i. e. for the state in which the value of U is the highest (in the situation shown in Fig. 2.4 this is the state u_3). If a system was initially in a uniform state that does not correspond to an absolute minimum of F, then, by creating a sufficiently large perturbation, it can be forced to go to the most stable state, corresponding to the deepest (absolute) minimum of F. Figure 2.8 shows the evolution of small (dot-and-dash) and large (solid line) perturbations of the uniform state $u = u_3$.

There is a profound similarity between the phenomena considered here and first-order phase transitions in equilibrium physical systems. Proceeding from this analogy, uniform distributions that correspond to minima of F are frequently called *phases* of such a system. A phase is called *metastable* if it corresponds to a local, but not the deepest, minimum of F. Such phase is unstable with respect to sufficiently strong perturbations. If a large nucleus of a stable phase is created inside it, this nucleus starts to grow (Fig. 2.8) and produces two spreading trigger waves that perform a transition into the most stable uniform steady state. A *critical nucleus* is, thus, a minimal perturbation that is sufficient to initiate the transition of an entire medium from a metastable to a stable uniform state. The size of such a nucleus is determined by the competition of two factors. Creation of a nucleus with u close to u_3 is favorable because this diminishes the first (potential) term in (2.3.2). On the other hand, in the presence of a nucleus, the distribution is no longer uniform and therefore the second term proportional to the square of grad u makes a positive contribution F. For a critical nucleus the two contributions cancel one another.

Note that a critical nucleus is given by a stationary unstable solution of equation (2.3.1).

In the one-dimensional case, the profile of a critical nucleus centered at $x = 0$ is a solution of the equation

$$D\frac{d^2u}{dx^2} + f(u) = 0 \tag{2.3.5}$$

with the boundary conditions

$$\frac{du}{dx} = 0 \quad \text{at} \quad x = 0 ,$$
$$u \to u_1 \quad \text{for} \quad |x| \to \infty \tag{2.3.6}$$

(we assume that the state u_1 is metastable).

Multiplying (2.3.6) by du/dx, integrating over x from 0 to ∞, and using the boundary conditions (2.3.7), we obtain an equation

$$\int_{u_1}^{u^*} f(u) \, du = 0 \tag{2.3.7}$$

that determines the value u^* of u in the center of a critical nucleus, i.e. at $x = 0$. Note that u^* is always smaller than u_3.

A special situation is created when both minima of F that corespond to the uniform distributions u_1 and u_3 have the same depth: $U(u_1) = U(u_3)$, or

$$\int_{u_1}^{u_3} f(u) \, du = 0 . \tag{2.3.8}$$

In this case, in an infinite medium, stationary coexistence of the two phases separated by a flat interface layer is possible.

This distribution corresponds to a higher value of F, since it includes an inhomogenity, namely the interface layer. Nevertheless, for an infinite medium, we cannot decrease F by pushing the interface out from the medium. Therefore, such a stationary nonuniform distribution has neutral stability.

If the dimensions of a medium are finite, the nonuniform distributions are unstable even when condition (2.3.5) is satisfied. An interface is driven to the medium border and, when it is finally expelled, the uniform state is recovered.

Note that, from a practical point of view, a medium can be considered infinite when its dimensions are large compared to the width of an interface layer. When condition (2.3.5) holds, such a large medium breaks down into the domains occupied respectively by the two different phases.

To conclude this section, we formulate some quantitative results concerning two- and three-dimensional media.

Obviously, the solution for a one-dimensional trigger wave constructed in Sect. 1.2 also describes trigger waves with flat fronts in two- or three-dimensional media. But what are the laws of motion for the waves with curved fronts?

Consider first the waves in a two-dimensional medium ($d = 2$). Any wave with a curved front can be divided into small segments approximated by circular arcs. Let us find the law of expansion of a circular front whose radius R is large compared to the width l of a transition layer.

In a polar coordinate system, taking into account the axial symmetry of the problem, (2.1.13) reduces to

$$\frac{\partial u}{\partial t} = f(u) + \frac{D}{r}\frac{\partial u}{\partial r} + D\frac{\partial^2 u}{\partial r^2} \ . \tag{2.3.9}$$

We can now remark that the derivative $\partial u / \partial r$ is actually vanishingly small everywhere except within a narrow transition layer of width l near $r = R$. Since we assume that $R \gg l$, we can approximately replace r by R in the factor (D/r) in (2.3.9).

Denote by $c(R)$ the propagation velocity of a circular trigger front with curvature radius R. Under our approximations, this front is described by a solution

$$u = u(\xi), \quad \xi = r - c(R)t \ , \tag{2.3.10}$$

of the differential equation

$$-c(R)u' = f(u) + (D/R)u' + Du'' \ , \tag{2.3.11}$$

satisfiying the boundary conditions $u \to u_1$ as $\xi \to \infty$ and $u \to u_3$ as $\xi \to -\infty$.

Rewriting (2.3.11) as

$$-[c(R) + (D/R)]u' = f(u) + Du'' \ , \tag{2.3.12}$$

we note that it would coincide with (2.2.3) for a flat trigger wave if we put $c = c(R) + D/R$. The boundary conditions are also identical. Therefore, if we know the propagation velocity c of a flat trigger wave, the propagation velocity $c(R)$ of a trigger front with curvature radius R is given by

$$c(R) = c - D/R \ . \tag{2.3.13}$$

Thus, a convex front moves more slowly than a flat one. If R is sufficiently small, $c(R)$ can become negative. Fronts with large curvatures do not propagate, but retreat. The radius of a stationary front that neither expands nor contracts is

$$R_{cr} = D/c, \quad d = 2 \ . \tag{2.3.14}$$

A circular domain with this radius represents a critical nucleus.

The above considerations are valid only for smooth fronts, with a local curvature radius much larger than the width l of a transition layer. Therefore, expression (2.3.14) can be used only in situations where the critical nucleus is sufficiently large, i.e. if $R_{cr} \gg l$. This condition is satisfied if the propagation velocity c of a flat front is small.

Although (2.3.13) was derived for a front which is convex into the region of a metastable phase, it remains valid for concave fronts. If we introduce the local curvature $K = 1/R$ of a front segment and assume it is positive for convex and negative for concave fronts, the propagation velocity of a front segment can be written as

$$c(K) = c - DK .\tag{2.3.15}$$

In problems of flame propagation, dependence (2.3.15) was found by *Markstein* [2.9]. Note that it ensures the stability of a flat front: if the front line is locally distorted, the convex portions of it will move more slowly than the concave ones and the distortion will be smoothed out.

The same arguments can be used for trigger waves in three-dimensional media. Every small portion of a front surface is then characterized by two principal curvatures K_1 and K_2. The propagation velocity is

$$c(K_1, K_2) = c - D(K_1 + K_2) .\tag{2.3.16}$$

A critical nucleus has the radius

$$R_{cr} = 2D/c, \quad d = 3 .\tag{2.3.17}$$

Expression (2.3.17) is valid only if $R_{cr} \gg l$, i.e. in media with sufficiently slow flat fronts.

2.4 Waves of Transition from an Unstable State

To conclude this chapter, we briefly discuss the properties of waves that perform transitions from unstable stationary states. At first glance, one can doubt why this case deserves special consideration. Indeed, some small fluctuations are usually present and they will destroy an unstable state even before the arrival of a transition wave. There is, however, an important class of problems, that involve spreading of populations, where such fluctuationary state can be maintained for indefinitely long time.

Suppose, for example, that we have some bacteria which can reproduce by fission or replication. Then the number of new bacteria born per unit time per unit volume is proportional to the number n of the bacteria already present in this volume element. Moreover, we can assume that the reproduction rate is directly proportional to the food density m. By taking into account diffusion of bacteria, we find that their population density obeys the equation

$$\frac{\partial n}{\partial t} = \alpha m n + D \Delta n .\tag{2.4.1}$$

The amount of food available is related to the number of consuming bacteria. In the simplest approximation, this relationship can be taken in the form

$$m = \begin{cases} m_0 - \beta n & n < n_0 , \\ 0 & n > n_0 , \end{cases}\tag{2.4.2}$$

where $n_0 = m_0/\beta$.

This system has two stationary uniform states, $n = 0$ and $n = n_0$. The first ($n = 0$) corresponds to the absence of a population. In the second state the population growth is prevented by exhaustion of the food supply.

The first state is unstable with respect to any small perturbations. Nevertheless, fluctuations cannot spontaneously arise in the absence of a population. Therefore such a state can be maintained indefinitely long, so long as the medium is not infected by bacteria.

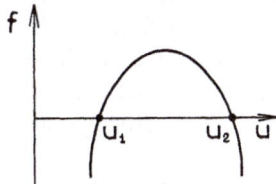

Fig. 2.9. Function $f(u)$ for a wave of transition from an unstable state

This biological model represents a special case of a wide class of problems with transitions from unstable states. Let us consider a medium described by an equation

$$\dot{u} = f(u) + D\Delta u \tag{2.4.3}$$

where the nonlinear function $f(u)$ has two roots, u_1 and u_2, corresponding to an unstable and a stable stationary states, respectively (see Fig. 2.9).

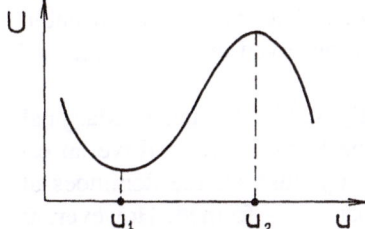

Fig. 2.10. Potential $U(u)$ for a wave of transition from an unstable state

The mathematical theory of waves that perform a transition from an unstable stationary state was constructed in 1937 by *Kolmogorov, Petrovskii*, and *Piskunov* [2.10]. A wave of transition from u_1 to u_2 is given by a special solution

$$u = u(\xi), \quad \xi = X - ct \tag{2.4.4}$$

of (2.4.3) that satisfies the boundary conditions

$$u \to u_2 \quad \text{for} \quad \xi \to -\infty,$$
$$u \to u_1 \quad \text{for} \quad \xi \to +\infty. \tag{2.4.5}$$

The wave profile $u(\xi)$ is determined by (2.2.3). We can again use a mechanical analogy, interpreting it as an equation of motion of a particle in the potential $U(u)$ defined by (2.2.4). However, the form of this potential is now different: $U(u)$ has a maximum at u_2 and a minimum at u_1 (Fig. 2.10).

In contrast to the case of trigger waves in bistable media, boundary conditions (2.2.4) do not specify a unique solution. There is a family of solutions that correspond to transition waves moving at different velocities. All possible propagation velocities are bounded from below. As we will soon see, however, the slowest transition wave propagating with a minimal velocity plays a special role.

In the framework of a mechanical analogy, boundary conditions (2.4.5) mean that we should release a particle at a time $\xi \to -\infty$ from a point u_2 where the potential has its maximum. Furthermore, we require that this particle arrives at the minimum point u_1 only at a time $\xi \to +\infty$. This is possible if the "viscous friction coefficient" c in (2.2.5) is not too small. Otherwise, the particle would arrive at u_1 in a finite time with nonvanishing velocity. Then it would pass this point and continue its motion in the negative direction. The minimal value of the friction coefficient that is sufficient to stop the particle at the point u_1 determines the minimal propagation velocity c_{min} of transition waves.

When propagation velocity c is large, we can neglect the diffusion term in (2.2.3), so that it reduces to

$$-cu' = f(u) \ . \tag{2.4.6}$$

Hence, propagation of a *fast* transition wave does not involve interaction between the neighboring elements of a medium. According to (2.4.5), this wave is described by a function

$$u = u^{(0)}(t - x/c) \tag{2.4.7}$$

where $u^{(0)}(t)$ is a solution of the dynamical equation (2.1.1) for an *isolated* unstable element. In effect, this is a fictitious wave that can be observed even in a chain of disconnected elements.

Consider toppling in a spaced array of marginally stable dominoes. Marginal stability means that any vanishingly small perturbation is sufficient to drive an element (a domino) from its stationary state. If we gently push all the dominoes at once, they fall down synchronously and no spatial pattern is formed. However, if we push the dominoes one after another, with a small delay, they will fall in turn and an illusion of a propagating wave will be created. This "propagation" does not involve interactions between the neighboring dominoes (so long as the delay is small compared with the time interval required by a falling domino to strike its neighbor). The smaller the delay, the higher the propagation velocity c.

Propagation of a fast transition wave is of a similar nature. If we apply small identical perturbations to a chain of unstable elements described by (2.1.1), but do so with a constant shift in time from one element in the chain to the next, the result is a transition wave propagating along the chain. Evolution of all elements is then identical but initial phases of this evolution are different. Smaller gradients of the initial phase correspond to higher propagation velocities.

On the other hand, for slow transition waves the phase gradients are larger and diffusion effects can no longer be neglected. Diffusion leads to spreading of perturbations to the neighboring elements, which drives them from a stationary state even in the absence of any direct external action.

The slowest possible transition wave (with a propagation velocity c_{min}) corresponds to *natural* propagation. Suppose an external perturbation is applied only at one end of the chain, to initiate the transition wave. Diffusion transfers the perturbation forward and causes the consecutive elements of the chain to leave the unstable stationary state. Hence, in all situations when the initial perturbation is strictly *localized*, the slowest possible transition wave is generated. This will be the case, for instance, in a biological model (2.4.1) with spreading infection.

Let us analyze further a relationship between trigger waves in bistable media and transition waves in an unstable medium. Suppose we take a function $f(u)$ for bistable medium (Fig. 2.1) and start to change it in such a manner that the plot of this function shifts upwards, retaining its form. Then points u_1 and u_2 move closer one to another and finally merge, so that we come to the situation shown in Fig. 2.9. The natural propagation velocity c_{min} of a transition wave in the resulting unstable medium coincides with the limit of the propagation velocity c_0 of trigger waves when these two zeros of the function $f(u)$ merge together.

We assumed above that the diffusion coefficient D is constant. However, there are some important problems where D depends on u, so that (2.4.3) is replaced by

$$\dot{u} = f(u) + \mathrm{div}\left(D(u)\mathrm{grad}\,u\right) . \tag{2.4.8}$$

Pattern formation in such unstable media is discussed by *Akhromeyeva* et al. [2.11].

3. Excitable Media

An element of an excitable medium returns to its initial state of rest after a burst of activity initiated by a supercritical external perturbation. The role of such a perturbation can be played by a diffusional flow from the neighboring elements of the medium. This results in propagation of a travelling excitation pulse.

In contrast to trigger waves in bistable media, an excitable medium goes back to its initial state after propagation of an excitation wave. Therefore, travelling waves can pass many times, one after another, through the same region. This property greatly enlarges the variety of possible wave patterns.

If we start with the initial condition in the form of a broken plane wave, under certain conditions the wave tip begins to sprout, simultaneously curling, and finally gives rise to a spiral wave. Spiral waves in excitable media are extremely stable wave patterns; their shape and rotation frequency are uniquely determined by the properties of a given medium.

When the properties of a medium vary in time or in space, the centers of spiral waves do not stay at fixed positions. Instead, spiral waves drift along in certain directions or perform more complicated meandering motion. Meandering of spiral waves can also be observed in some stationary uniform media, where the steady rotation of a wave is inherently unstable.

In three dimensions, excitation waves can produce very complicated vortex patterns, such as simple or twisted scrolls, rings, knots etc. Investigation of dynamical properties of such patterns is a challenging problem in the theory of active media.

3.1 Travelling Pulses

There is a close relationship between the phenomena in bistable and excitable media. In effect, we can transform a bistable medium into an excitable one by introducing a mechanism that restores the initial state of elements after passage of a trigger wave.

Suppose we have a combustion process with a heat production rate q which depends not only on temperature θ but also on the concentration v of a certain substance (the *inhibitor*), so that $q = q(\theta, v)$. We assume that the presence of an inhibitor decreases heat production.

At a fixed inhibitor concentration, thermal wave propagation is described by the equation

$$\frac{\partial \theta}{\partial t} = f(\theta, v) + \chi \frac{\partial^2 \theta}{\partial x^2} , \qquad (3.1.1)$$

where

$$f(\theta, v) = -\gamma(\theta - \theta_1) + q(\theta, v)/C . \tag{3.1.2}$$

As it was shown in Sect. 2.2, the direction of transition in a trigger wave and the wave velocity depend on the parameter

$$A(v) = \int_{\theta_1}^{\theta_3} f(\theta, v)d\theta . \tag{3.1.3}$$

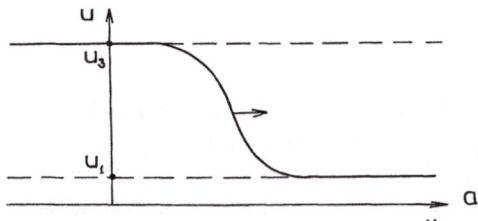 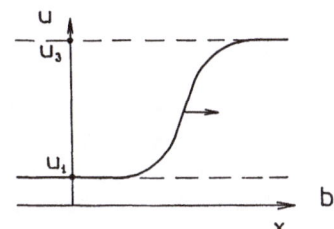

Fig. 3.1a,b. Trigger waves for (a) $A > 0$ and (b) $A < 0$

If A is positive, passage of a trigger wave results in a transition from the cold state θ_1 to the hot state θ_3. Hence, this is an *ignition* wave. On the other hand, if $A < 0$, the direction of transition is reversed and we have a wave of *quenching* (Fig. 3.1).

Since parameter A depends on v, propagation of a thermal wave is controlled by the inhibitor concentration. By varying v we can change its velocity and even reverse the propagation direction. When the inhibitor concentration is low, this is an ignition wave. At higher inhibitor concentrations it represents a wave of quenching.

Suppose now that the inhibitor is produced in the combustion process as a reaction by-product. Production of the inhibitor should be accompanied by its decay or dissipation; otherwise it would accumulate in the medium. An equation for the inhibitor concentration which takes into account both these effects is

$$\frac{\partial v}{\partial t} = -\frac{1}{\tau}(v - \bar{v}(\theta)) . \tag{3.1.4}$$

Here $\bar{v}(\theta)$ is an equilibrium inhibitor concentration that sets in if we fix temperature θ; it is a monotonously increasing function of θ.

In our example we assume that the characteristic time τ of the inhibitor variation is large compared with the time-scale of temperature variations. If the relationship between these two characteristic times were reversed, the inhibitor concentration would adjust adiabatically to the momentary temperature. But substitution of the equilibrium concentration $v = \bar{v}(\theta)$ into (3.1.1) would lead us back to the one-component model of a bistable medium that has already been discussed.

Moreover, we assume that diffusion of the inhibitor is very slow and can be neglected. The effects of fast inhibitor diffusion will be discussed in Chap. 5.

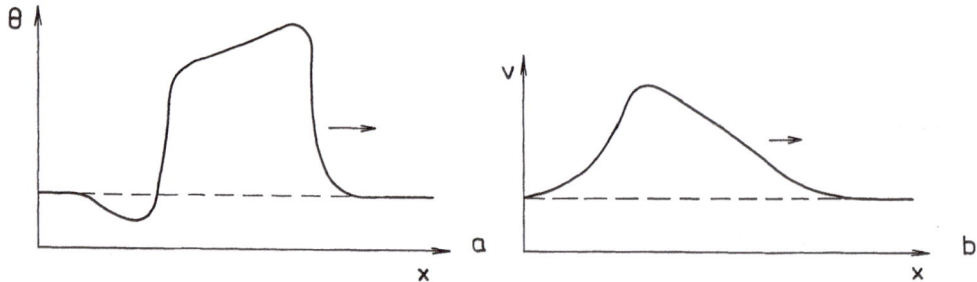

Fig. 3.2a,b. Distributions (a) of temperature (or of an activator concentration) and (b) of inhibitor concentration in a solitary travelling pulse

A solitary travelling pulse in the medium described by (3.1.1) and (3.1.4) has the form shown in Fig. 3.2. It consists of a sharp front and rear side, where rapid changes of temperature occur, along with a top and a tail where variations of temperature are much slower. On the other hand, the inhibitor concentration has no rapid variations: it slowly increases while the top of a pulse is passing and then slowly diminishes until the initial concentration is reached.

The front and the rear side of a pulse are, in effect, two waves of ignition and quenching that move one after another at the same speed. The first wave propagates on a background of low inhibitor concentration typical for the rest state. After its passage, fire sets in and temperature sharply increases. Then the inhibitor concentration starts to grow and, reaching a certain critical value, quenches the fire. The temperature rapidly drops, production of the inhibitor is stopped, and its concentration slowly decreases. After a while, the medium is brought back to the initial state. Now we can ignite the medium again.

Solitary pulse propagation with subsequent restoration of an initial state can also be realized by another mechanism that does not involve an inhibitor. Imagine a fire storm in a wild steppe. When the flame passes all the dry grass is burned out, but later it starts to grow up again and, after some long period of time, the initial state of this medium is restored. Then another fire storm can occur.

Waves in excitable media should not be always related to thermal phenomena. In effect, combustion is only an example of a self-accelerating process. Instead of that example, one can consider the reproduction of certain molecules or biological microorganisms, where an activation effect can be achieved by a diffusional flow of reproducing molecules or bacteria.

Generally, a mechanism should be present that stops the avalanche caused by a self-accelerating process (this can be related either to exhaustion of certain resources or to the by-production of an "inhibitor"). At the final stage, some restoration mechanism should come into operation which brings the medium back to the initial state.

The best known example of a realistic excitable (and oscillatory) medium is that of a chemical active medium with the Belousov-Zhabotinskii (BZ) reaction. This reaction was discovered in 1951 by *Belousov* (who was able to publish the first short communication [3.1] on this subject only in 1959) and was later investigated by *Zhabotinskii* [3.2, 3]. The reaction proceeds in a dilute solution and its thermal effects are negligible.

A detailed description of the BZ reaction can be found in a review by *Tyson* [3.4]. This reaction includes an autocatalytic stage that results in the reproduction of certain active molecules. For a special choice of the parameters, the quiescent state of the solution is stable under small perturbations, and under normal conditions the reaction can be triggered only by a diffusional flow of the activator molecules. In the course of this reaction, inhibitor molecules are produced that slow down the reaction and finally stop it. Later the inhibitor disappears (since it takes part in other reactions as well). The BZ reaction is irreversible, and hence each one of its bursts results in the consumption of a portion of the chemicals stored in the solution. However, a rich solution can support tens of individual bursts without any essential change in its composition. Therefore, it can be assumed as an approximation that the initial state of this medium is restored after every burst.

Spatial patterns in a chemical medium with the BZ reaction are very convenient for an experimental study. This reaction occurs under common laboratory conditions in a thin layer of a liquid solution. Propagation of excitation waves is manifested by local changes of the solution color. All processes are sufficiently slow (for instance, the excitation pulse moves at a speed of about several millimeters per minute) and they can be observed by the unaided eye (the characteristic size of typical details is about 1 mm).

The complete chemical scheme of the Belousov-Zhabotinskii reaction is very complicated: it involves more than 20 intermediate stages. In a simplified form, taking into account differences in the kinetic rates for various stages, the BZ reaction can be described by the following model with two effective components (which was introduced by *Tyson* and *Fife* [3.5] who used the results of the analysis by *Field*, *Koros*, and *Noyes* [3.6]):

$$\frac{\partial u}{\partial t} = u(1 - u) - \frac{v(u - a)}{u + a} + D_u \frac{\partial^2 u}{\partial x^2} \, ,$$
$$\frac{\partial v}{\partial t} = -\frac{1}{\tau}(v - bu) + D_v \frac{\partial^2 v}{\partial x^2} \, .$$

$$(3.1.5)$$

Here u represents the normalized concentration of the activator molecules. The second variable v can be formally interpreted as an inhibitor concentration (this interpretation is not quite correct from a chemical point of view). Equations (3.1.5) are written in dimensionless units, and a and b are positive parameters ($a \ll 1$). The inhibitor evolution is slow ($\tau \gg 1$).

Other important examples of excitable media are of physiological origin. For instance, propagation of an excitation pulse along a nerve fiber can be approximately described by a model (see the review by *Scott* [3.7])

$$C\frac{\partial U}{\partial t} = I_{\text{Na}}(U) - I_{\text{K}}(U, \sigma) + \frac{1}{R}\frac{\partial^2 U}{\partial x^2} \, ,$$
$$\frac{\partial \sigma}{\partial t} = -\frac{1}{\tau}\left[\sigma - \bar{\sigma}(U)\right] \, .$$

$$(3.1.6)$$

Here C is the specific membrane capacity per unit fiber length, R is the specific electrical resistance of axoplasma filling this fiber, I_{Na} and I_{K} are local sodium and potassium ion currents through the membrane, $\bar{\sigma}(U)$ is the equilibrium value of the electrical conductivity σ for a potassium current at a fixed voltage difference U on a membrane. Evolution of the potassium conductivity σ is a slow process characterized by a large time-scale τ.

The heart tissue is formed by tangled fibers that are also described by (3.1.6). When such fibers touch, electrical contact is established. Therefore, in the continuum approximation this tissue can be considered as a distributed two- or three-dimensional medium, depending on its thickness.

Some further examples of excitable media are discussed by *Vasilev* et al. [3.8]. Below we consider a class of excitable media described by the equations

$$\frac{\partial u}{\partial t} = f(u, v) + D\Delta u \; ,$$

$$\varepsilon^{-1}\frac{\partial v}{\partial t} = -v + \bar{v}(u) \; ,$$

$$(3.1.7)$$

where $\bar{v}(u)$ is some monotonously increasing function and ε is a small parameter ($\varepsilon \ll 1$). The null-clines $f(u, v) = 0$ and $v = \bar{v}(u)$ intersect at a point that corresponds to the rest state of the medium (Fig. 3.3). In the following u is called the activator concentration and v is called the concentration of an inhibitor although the precise interpretation of these variables might depend on the particular problem.

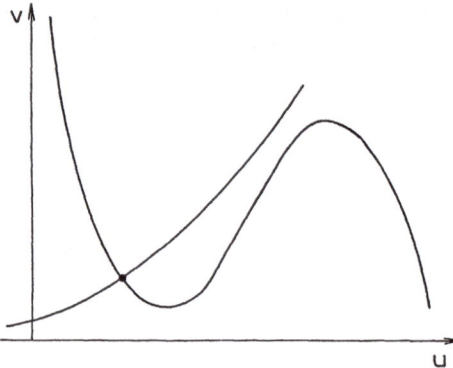

Fig. 3.3. Null-clines of excitable medium

First we give a qualitative description of travelling pulses, based on the papers by *Ortoleva* and *Ross* [3.9] and by *Ostrovskii* and *Yakhno* [3.10].

Note that v is a slow variable with a characteristic time-scale about $1/\varepsilon$. Therefore, in the front and the rearside regions of a travelling pulse, where u is rapidly changing, the variable v remains constant. Variations of u in the front and the rearside regions are described by the first equation of (3.1.7) where v is taken as a fixed parameter. This coincides with equation (2.1.13) for propagation of trigger waves.

Hence, the front of the travelling pulse represents a trigger wave. Its propagation velocity (and, consequently, the propagation velocity of the entire pulse) is determined by the first equation of (3.1.7) with a low inhibitor concentration $v = v_0$ found in the rest state. The rearside is also a trigger wave, but this time of the reverse transition, *that should follow after the front at the same velocity*. The latter condition determines the inhibitor concentration $v = v^-$ at which the reverse transition takes place in a steadily travelling pulse.

An interval between the front and the rearside corresponds to the top of a pulse. In this region the inhibitor concentration slowly increases from v_0 to v^- with a characteristic time ε^{-1}. The activator concentration u adjusts adiabatically to momentary values of v, and its diffusion is not important.

At the rearside of a pulse the activator concentration drops sharply. However, the inhibitor is much more inertial, and hence its concentration v can return from v^- to the initial value v_0 only after some time interval with a characteristic time ε^{-1}. This interval corresponds to the tail of a pulse.

Since the front and the rearside of a pulse are narrow (they have characteristic times of order unity), the total width of a pulse is mainly determined by its top and tail. The duration of a pulse is about ε^{-1}.

A rigorous solution to the problem of a solitary travelling pulse was given by *Casten* et al. [3.11] who used *singular perturbation theory* (a popular introduction to this theory can be found in the lectures by *Murray* [3.12]).

Let us introduce a moving reference frame $\xi = x - V_0 t$ where the pulse is resting. Then (3.1.7) reduces to a system of two ordinary differential equations

$$-V_0 u' = f(u,v) + D u'' ,$$
$$-(V_0/\varepsilon)v' = \bar{v}(u) - v ,$$

(3.1.8)

that should be solved with the conditions $u \to u_0$ and $v \to v_0$ as $\xi \to \pm\infty$, where (u_0, v_0) is the stationary uniform rest state of the medium. A nontrivial solution satisfying these conditions represents a separatrix loop of a stationary saddle point (u_0, v_0) in the phase plane. But such a special trajectory exists only at a definite value of the parameter V_0 in (3.1.8), which is the propagation velocity of a solitary pulse.

In the leading approximation, the propagation velocity V_0 of a solitary pulse coincides with the velocity c_0 of a trigger wave described by the first of equations (3.1.7) with $v = v_0$. By using a singular perturbation technique (see [3.11]), one can estimate the first correction to this result:

$$V_0 \approx c_0(1 - \xi_0 \varepsilon) ,$$

(3.1.9)

where ξ_0 is a numerical factor independent of ε.

Equations (3.1.7) also have solutions that describe a travelling periodic train of pulses. If the spatial period L of pulses in a train is much larger than the total width of a single pulse (including its tail), the train moves at the same velocity as a solitary pulse. However, for smaller periods L the train velocity decreases because the front of the next pulse in the train then moves on a background of the residual inhibitor concentration left by the tail of the previous pulse.

Formally, a periodic train of pulses is described by a special solution $u = u(\eta)$, $v = v(\eta)$, where $\eta = kx - \omega t$; $u(\eta + 2\pi) = u(\eta)$, $v(\eta + 2\pi) = v(\eta)$. It obeys a system of two ordinary differential equations

$$-\omega u' = f(u,v) + D k^2 u'' ,$$
$$-(\omega/\varepsilon)v' = \bar{v}(u) - v$$

(3.1.10)

and corresponds to a limit cycle of the system. The period of this limit cycle is

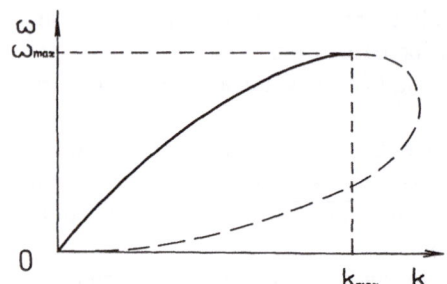

Fig. 3.4. The typical form of the dispersion relation for nonlinear periodic waves in excitable media. The dashed line corresponds to the unstable solution

determined by the values of paramters ω and k in (3.1.10). If we require that the period is equal to 2π, we impose a certain relationship on ω and k, leading to a dependence $\omega = \Omega(k, \varepsilon)$. This dependence can be considered as a *dispersion relation* for nonlinear periodic waves (see *Mikhailov* and *Krinsky* [3.13]).

A typical form of the dispersion relation is shown in Fig. 3.4. The right branch (dash line) of the dispersion curve, which describes slow waves, corresponds to unstable solutions. Generally, the instability area extends into the region of fast waves (cf. [3.14, 15]). However, to date there have been no analytical calculations of the stability border. Computer simulations indicate that it can lie close to the boundary between the regions of fast and slow waves. Therefore, it is often assumed that the entire left branch of the dispersion curve corresponds to the set of stable solutions.

Hence, there is a minimal spatial period $L_{min} = 2\pi/k_{max}$ of a stable train of pulses, which is about the width of a single travelling pulse. It corresponds to the maximal possible frequency of pulses $\omega_{max} \sim \varepsilon^{1/2}$.

As already mentioned in Sect. 2.2, even analytical calculation of the propagation speed of a trigger wave in simple bistable media is usually impossible. The difficulties are still greater in the case of travelling pulses in excitable media. Commonly, one has to resort here to numeric calculations. An important exception, however, is provided by the *Rinzel-Keller model* [3.16], given by the equation

$$\frac{\partial u}{\partial t} = -u + \bar{u}H(u - u_0) - v + \frac{\partial^2 u}{\partial x^2} ,$$
$$\varepsilon^{-1}\frac{\partial v}{\partial t} = u ,$$
(3.1.11)

where the step-like Heaviside function is defined as $H(z) = 0$ for $z < 0$ and $H(z) = 1$ for $z \geq 0$.

The dispersion relation for nonlinear waves in the Rinzel-Keller model was analytically calculated by *Mikhailov* and *Krinsky* [3.13] (see also [3.17]). The maximal frequency of a periodic pulse train in this model is

$$\omega_{max} = \frac{\pi\varepsilon}{\text{arctanh}\left[\sqrt{3\varepsilon}/2\mu_0\right]} ,$$
(3.1.12)

where $\mu_0 = \frac{1}{2} - (\bar{u}_0/u)$. No self-supported propagation of pulses is possible if $\mu_0 < \sqrt{3\varepsilon}/2$. The maximal wavenumber is

$$k_{max} = \frac{\pi\sqrt{\varepsilon/3}}{\operatorname{arctanh}\left[\sqrt{3\varepsilon}/2\mu_0\right]} \tag{3.1.13}$$

When condition $\mu_0 \ll 1$ holds, the dispersion relation can be approximated in an implicit form by

$$\omega = 4k\mu_0 \tanh(\pi\nu) - 3\nu k^2 , \tag{3.1.14}$$

where $\nu = \varepsilon/\omega$.

Note that a solution for the periodic sequence of pulses with a spatial period L describes, at the same time, propagation of a solitary pulse along a circle of perimeter L.

In a two-dimensional medium, a solitary travelling pulse becomes a solitary excitation wave. Below we analyze the dependence of propagation velocity V of such a wave on the local curvature K of its front.

In order to find the dependence of V on K, let us first introduce a local polar coordinate system, such that its center coincides with the center of curvature of a given small segment of a front. If the curvature radius $R = 1/K$ is much larger than the width of a wave (i. e. the width of a travelling pulse), we can write approximately

$$\frac{\partial u}{\partial t} = f(u, v) + \frac{D}{R}\frac{\partial u}{\partial r} + D\frac{\partial^2 u}{\partial r^2} ,$$
$$\varepsilon^{-1}\frac{\partial v}{\partial t} = \bar{v}(u) - u . \tag{3.1.15}$$

The concentric wave propagation at speed V is described by a solution of the form $u = u(\xi), v = v(\xi)$ where $\xi = r - Vt$, satisfying equations

$$-(V + DK)u' = f(u, v) + Du'' ,$$
$$-(V/\varepsilon)v' = \bar{v}(u) - u . \tag{3.1.16}$$

These reduce to equations (3.1.8) for propagation of a solitary pulse if we put $V^* = V + DK$ and replace ε by $\varepsilon^* = (V + DK)\varepsilon/V$ in (3.1.8). Thus, the dependence of V^* on ε^* is given by the relationship (3.1.9) found above for a solitary pulse. Substituting expressions for V^* and ε^* into (3.1.9) and solving the resulting equation, we find after *Zykov* [3.18] that

$$V/c_0 = \tfrac{1}{2}(1 - \xi_0\varepsilon - DK/c_0)$$
$$\pm \tfrac{1}{2}\sqrt{(1 - \xi_0\varepsilon - DK/c_0)^2 - 4\xi_0\varepsilon DK/c_0} . \tag{3.1.17}$$

This dependence is shown in Fig. 3.5. No wave can propagate with a curvature larger than

$$K^* = (c_0/D)\left[1 - 2\sqrt{\xi_0\varepsilon}\right] . \tag{3.1.18}$$

For $K < K^*$, there are two branches, the lower of which (dashed line) is always unstable. The computational data (see a monograph by *Zykov* [3.19] indicates that

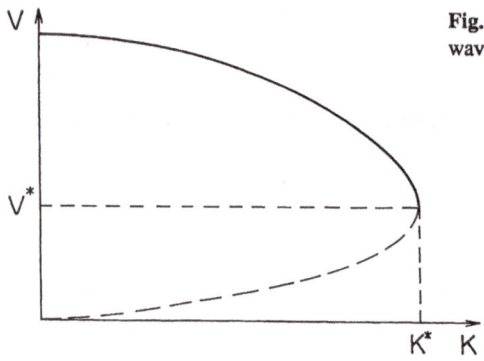

Fig. 3.5. Dependence of the propagation velocity V on the wave curvature K in excitable media

waves in the upper branch remain stable even when the curvatures of their fronts come very close to K^*.

It is instructive to compare the properties of propagating curved trigger waves in two-dimensional bistable media and the above results for propagation of curved excitation waves in excitable media. As pointed out in Sect. 2.3, the propagation velocity of a trigger wave is related to its curvature by a simple linear law $c = c_0 - DK$. At high enough curvatures, a front can even stop or reverse the direction of its motion. For excitation waves this is impossible. When the curvature of their front comes close to K^*, such waves lose stability and disappear. An excitation wave cannot stop or reverse the direction of its propagation. Waves with a curvature near K^* propagate at a velocity $V^* \approx (\xi_0 \epsilon)^{1/2}$.

Although our analysis was limited to a particular class of reaction-diffusion equations given by (3.1.7), many of its results remain valid for other two-component models of excitable media as well (see [3.8]).

3.2 Cellular Automaton Models

In this section we consider the evolution of wave patterns in discrete networks of connected excitable elements. Instead of using differential equations, we construct the model in terms of *cellular automata* with discrete time and a finite number of possible states. The state of any such element at a given time step $n + 1$ is uniquely determined by the states in which this element and its neighbors in the network were found at the previous step n.

There are two principal reasons which motivate us to use such models.

Even if certain differential equations (e. g. equations (3.1.7)) that describe a given medium are known, it is usually impossible to find analytical solutions of these equations, and numeric computations need to be performed. If the problem consists in investigation of the wave processes related to formation and development of complex patterns in two- or three-dimensional media, computations become very tedious. In this situation (especially if we are interested mainly in qualitative results) we can abstain from a numeric integration of exact differential equations and turn to the analysis of much simpler systems represented by networks of cellular automata.

Simulations show that their dynamics reflect a large number of phenomena found in continuous excitable media.

Besides that, there are situations where cellular automata models are even better than the approximation of a continuous medium.

In 1946 *Wiener* and *Rosenblueth* [3.20] proposed a simple model that allows one to analyze various regimes of excitation propagation in a homogeneous neuron network or in heart tissue. According to this model, an excitable medium consists of elements that have three possible states, of rest, excitation, and refractoriness. Moved into an excited state, such an element stays there for a fixed time[1], then goes to the state of refractoriness and only after that returns to the state of rest. An element can undergo a transition into the excited state if one of the neighboring elements is excited; this transition can also be brought about by some external action. Transitions into the excited state are possible only from the state of rest.

In the framework of their model, Wiener and Rosenblueth described the phenomenon of a spiral wave circulation around a hole in the excitable medium. Later *Selfridge* [3.22] and *Balakhovskii* [3.23] showed that presence of a hole is not a necessary condition for the existence of spiral waves. *Krinsky* [2.24] demonstrated that spiral waves are produced when a wavefront breaks in a strongly inhomogeneous medium. Cellular automata were used as models of excitable media by *Reshodko* and *Bures* [3.25], *Mádore* and *Freedman* [3.26], *Winfree* et al. [3.27], *Gerhardt* et al. [3.86], *Markus* and *Hess* [3.87].

Below we consider a generalized Wiener-Rosenblueth model (proposed by *Mikhailov* and *Zykov* [3.28]) which takes into account the threshold property of activation and temporal summation of arriving signals.

Suppose that a two-dimensional network consists of elements enumerated by a pair of indexes, i and j. A state of any element is specified by two variables Φ_{ij}^n and u_{ij}^n; here the upper index n indicates the discrete time step. Transitions between the states obey the following set of rules:

$$\Phi_{ij}^{n+1} = \begin{cases} \Phi_{ij}^n + 1 \,, & \text{if } 0 < \Phi_{ij}^n < \tau_e + \tau_r \,, \\ 0 \,, & \text{if } \Phi_{ij}^n = \tau_e + \tau_r \,, \\ 0 \,, & \text{if } \Phi_{ij}^n = 0 \text{ and } u_{ij}^{n+1} < h \,, \\ 1 \,, & \text{if } \Phi_{ij}^n = 0 \text{ and } u_{ij}^{n+1} \geq h \,. \end{cases} \qquad (3.2.1)$$

Here Φ_{ij}^n is an integer *phase* of the element located in the site (i,j). A zero phase ($\Phi_{ij}^n = 0$) corresponds to the state of rest. If $0 < \Phi_{ij}^n \leq \tau_e$, an element is said to be in an excited state. A refractory state corresponds to the phase values $\tau_e < \Phi_{ij}^n \leq \tau_e + \tau_r$. According to (3.2.1), an element in the site (i,j) goes from the state of rest to an excited state if u_{ij}^{n+1} exceeds a threshold h. Afterwards this element performs a fixed sequence of transitions, each increasing its phase by one. When its phase becomes equal to $\tau_e + \tau_r$, an element returns to the initial state of rest.

The quantity u_{ij}^n is interpreted as an "activator concentration" in the site (i,j). An "activator" is produced by elements in the excited state; it also decays. These

---

[1] In [3.20] the duration of an excited state was assumed to be vanishingly small; a model with a finite excitation time was proposed later by *Rosenblueth* [3.21].

two effects are described by an updating rule

$$u_{ij}^{n+1} = gu_{ij}^n + \sum_{k,l} C(k,l)I_{i+k,j+l}^n ,$$ (3.2.2)

where

$$I_{ij}^n = \begin{cases} 1 , & \text{if } 0 < \Phi_{ij}^n \leq \tau_e , \\ 0 , & \text{if } \tau_e < \Phi_{ij}^n \leq \tau_e + \tau_r \text{ or } \Phi_{ij}^n = 0 . \end{cases}$$ (3.2.3)

Coefficients $C(k,l)$ determine the range of interaction; they vanish at large k and l. Usually it is sufficient to take into account only interactions between neighboring elements, i.e. we set

$$C(k,l) = \begin{cases} 1 , & \text{if } |k| \leq 1 \text{ and } |l| \leq 1 , \\ 0 , & \text{otherwise} . \end{cases}$$ (3.2.4)

The model given by (3.2.1–3) reduces to the original Wiener-Rosenblueth model if we take $h = 1, g = 0$ and choose to set $C(k,l)$ in the form (3.2.4). Then, any excited element can excite at time step n all the neighboring elements that are found in the state of rest in the preceding time step $n - 1$.

If $h > 1$ and $0 < g \leq 1$, an element waits until a threshold activator concentration h is reached which can occur either because of a temporal summation or because of a simultaneous flow of activator from several neighboring elements. Hence, h is interpreted as an excitation threshold, and $1/(1 - g)$ is the halflife of an activator.

With different relationships between its parameters this model displays a large variety of behaviors[2].

If $h = 3$, such a medium supports the propagation of a flat excitation wave (Fig. 3.6) moving at a velocity $V = 1$, whose front shifts by one lattice element

$n = 1$

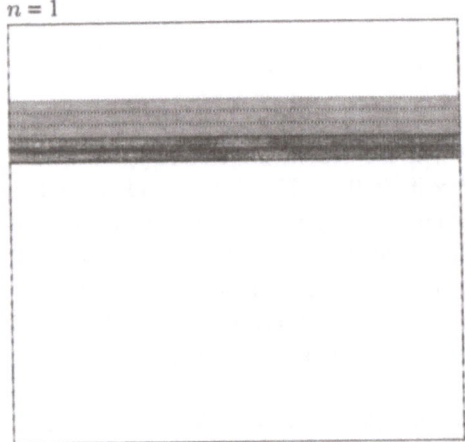

Fig. 3.6. Solitary flat wave in a cellular automaton model of an excitable medium. Dark cells correspond to the states of excitation, grey cells show the elements in the refractory states, blank cells indicate the elements in the state of rest

[2] The simulations which are described below were performed by *Dorjsurengiyn* [3.29] (see also [3.30]).

$n = 1$

$n = 15$

$n = 8$

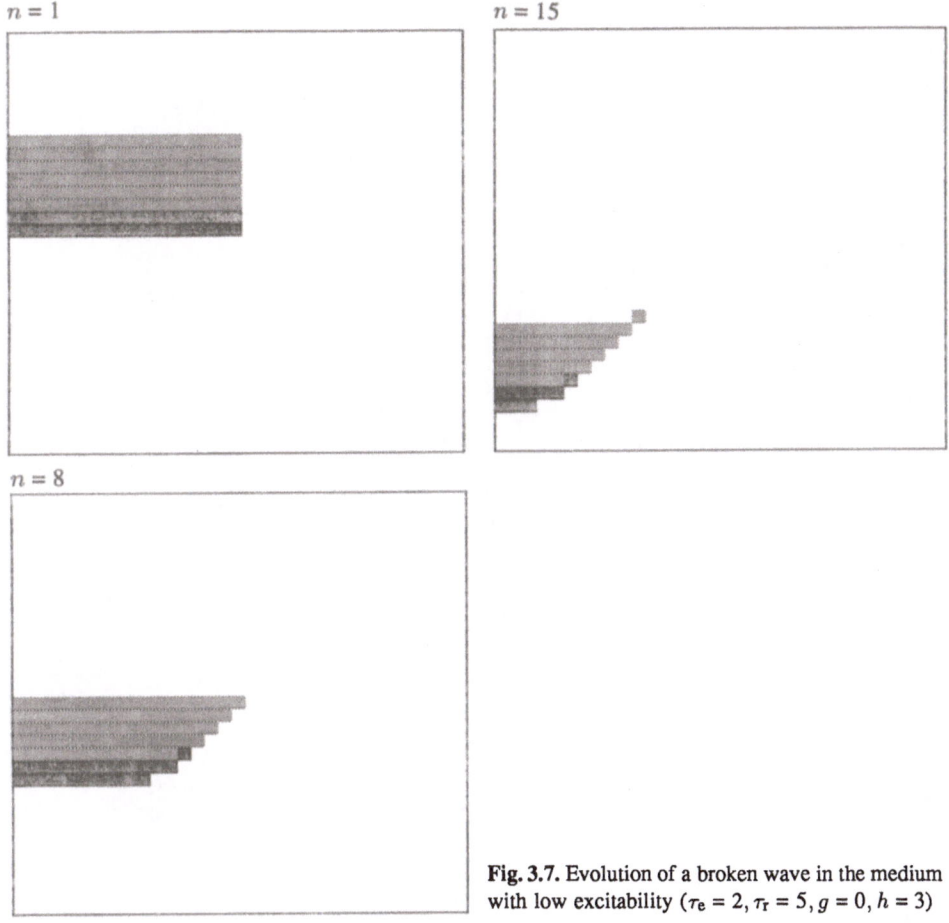

Fig. 3.7. Evolution of a broken wave in the medium with low excitability ($\tau_e = 2, \tau_r = 5, g = 0, h = 3$)

with every time step. This wave has a total width of $\tau_e + \tau_r$. For $h > 3$, flat waves propagate at smaller velocities $V \le \frac{1}{2}$, i. e. their fronts shift only at every second (or every third, etc.) step of time. When $h > 3$, in order to make propagation possible, the activator should not decay too fast, i. e. the coefficient g should not be too small.

Since the state of the medium before and after the passage of a wave is the same, nothing prevents us from taking the initial condition to be in the form of a broken half-wave (Fig. 3.7). What would be the subsequent evolution of such an initial distribution? It turns out that the outcome depends very essentially of the effective *excitability* of the medium, determined by the excitation threshold h, the duration of an excited state τ_e, and the value of the coefficient g (effective excitability increases with τ and g, and diminishes when the excitation threshold h is increased).

If the excitability is low, a half-wave contracts (Fig. 3.7) and the initial excitation disappears. By choosing special values of the parameters, one can achieve a situation in which a half-wave moves forward without either contracting or growing. For higher excitabilities, a half-wave grows (sprouts) at its free end and its progress is retarded there. As a consequence, a rotating *spiral wave* is produced (Fig. 3.8). When

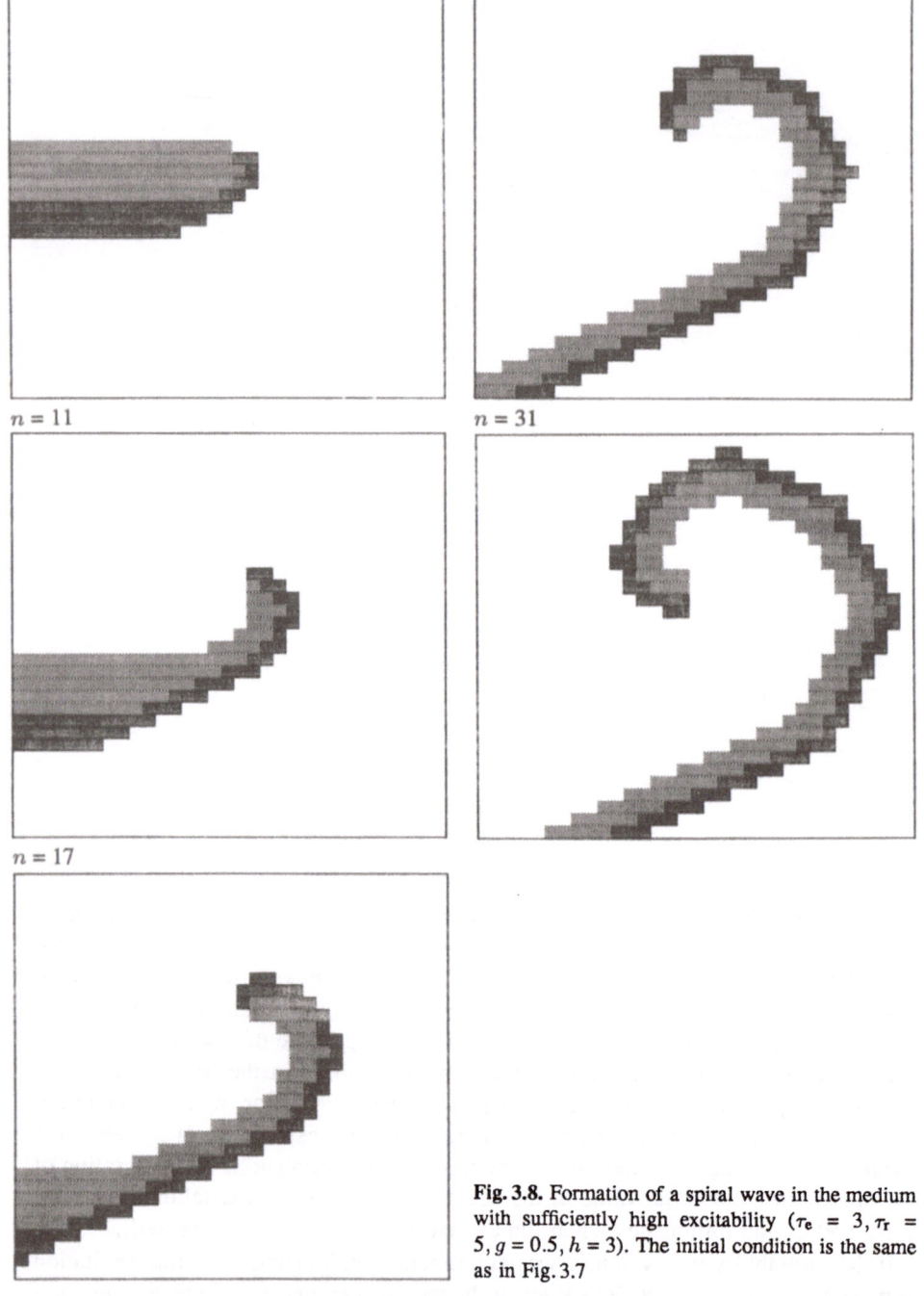

Fig. 3.8. Formation of a spiral wave in the medium with sufficiently high excitability ($\tau_e = 3, \tau_r = 5, g = 0.5, h = 3$). The initial condition is the same as in Fig. 3.7

$n = 24$

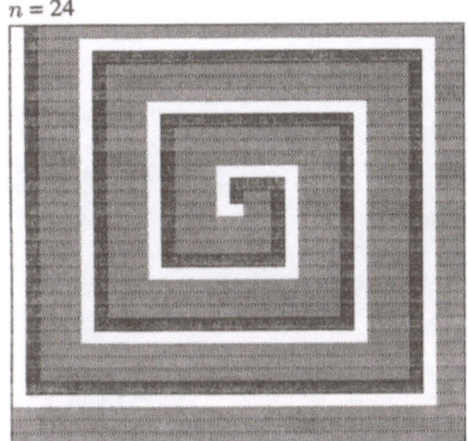

Fig. 3.9. Spiral wave in the Wiener-Rosenblueth model ($\tau_e = 1, \tau_r = 3, g = 0, h = 1$)

excitability is not too high, this spiral is rather sparsely spaced, i.e. the subsequent coils of a spiral are separated by large regions where elements are in the state of rest. The end point then performs a circular motion, rotating around a rest region (the *core* of the spiral wave), which can be very large. At still higher excitabilities, the spiral wave is curled more tightly and its core is smaller.

A spiral wave for the Wiener-Rosenblueth model, corresponding to the limit of an extremely high excitability, is shown in Fig. 3.9. We see that in this case the core of the spiral wave (i.e. the quiescent region in its center) is absent.

If we take the initial condition to be in the form of two half-waves with opposite directions (Fig. 3.10), its evolution results in the creation of a two-armed spiral. Note the characteristic variations in the central area and the periodic "touching" of waves there.

Our simulations show that colliding excitation waves annihilate. This is a general property of waves in excitable media. Indeed, when two such waves collide, excited elements are surrounded on both sides by elements in a state of refractoriness, and therefore they cannot transfer excitation to other elements of the medium.

Annihilation of colliding waves leads to a very important effect. If we have several periodic wave sources in the same medium, then a source which generates the sequence of waves with the highest frequency will suppress all other sources. Figure 3.11 shows the time evolution of a wave pattern from two sources with different generation periods T_1 and T_2. A point where two waves meet shifts closer after each collision to a source with a lower frequency, until its action is suppressed.

A local periodically acting source of waves (i.e. a *pacemaker*) can be created if we take a group of oscillatory elements with a phase dynamics that obeys a simple rule

$$\Phi_{ij}^{n+1} = \begin{cases} \Phi_{ij}^n + 1 , & \text{if } \Phi_{ij}^n < T - 1 , \\ 0 , & \text{if } \Phi_{ij}^n \geq T - 1 . \end{cases} \qquad (3.2.5)$$

Evidently, any such element will, independently of its neighbors, perform oscillations with a period T. If we place a sufficiently large group of such elements into

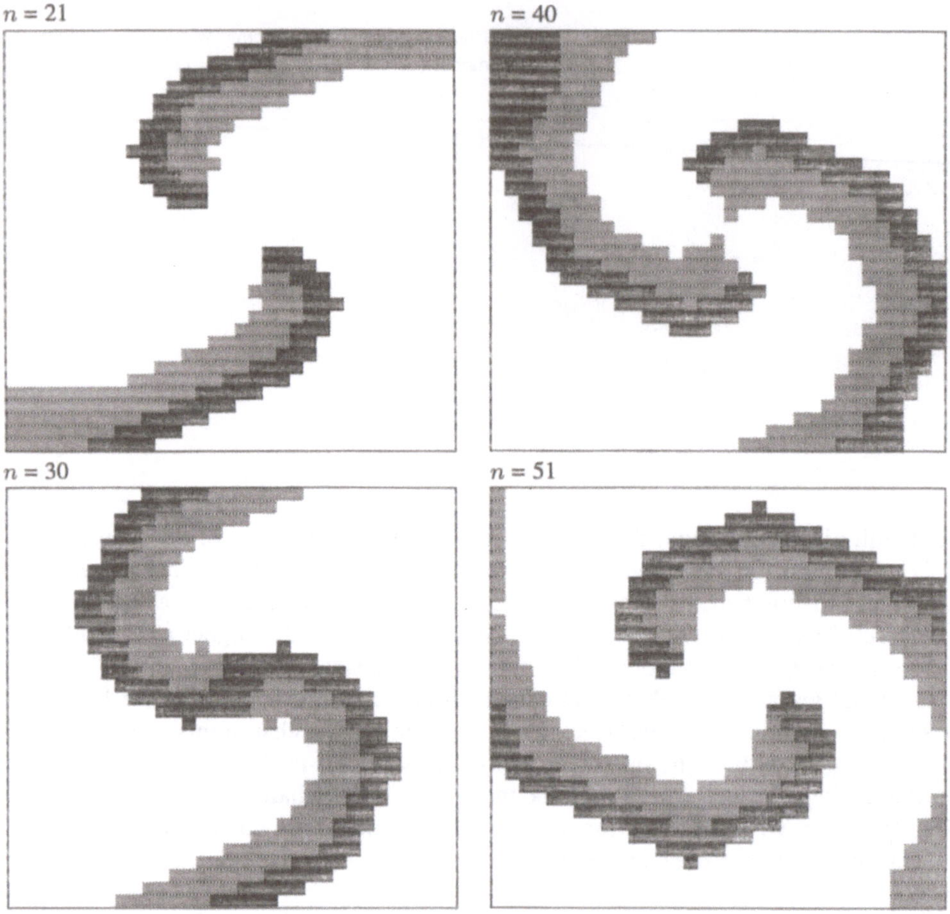

$n = 21$

$n = 40$

$n = 30$

$n = 51$

Fig. 3.10. Two-armed spiral wave ($\tau_e = 5, \tau_r = 7, g = 0, h = 3$)

$n = 65$

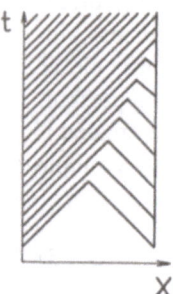

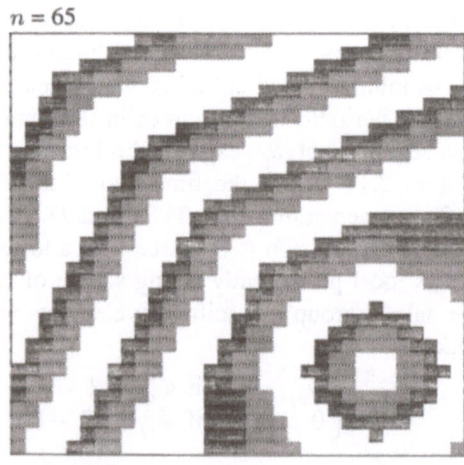

Fig. 3.11. Time evolution of colliding waves from two sources with different generation periods

Fig. 3.12. Pacemaker ($\tau_e = 3, \tau_r = 4, g = 0.5, h = 3, T = 15$)

an excitable medium, this group will function as a source of concentric spreading excitation waves (Fig. 3.12).

Note that the effective generation period of a pacemaker coincides with T only if $T \geq \tau_e + \tau_r + 1$. Otherwise some beats are skipped, i. e. they do not produce spreading circular waves, since the excited central group of elements finds itself surrounded by elements in the state of refractoriness. In this case a pacemaker generates waves at higher periods $2T, 3T$, etc.

Figure 3.13 shows how the activity of a low-frequency pacemaker is suppressed by another pacemaker with a higher generation frequency.

We see that pacemakers appear as a result of an inhomogenity, i. e. their existence is related to the presence of an oscillatory group of elements in an otherwise homogeneous medium. In contrast to this, spiral waves do not require any inhomogenities. The position of a spiral wave center is determined only by the initial conditions that lead to its creation.

All spiral waves in a given medium have the same rotation frequency. Therefore, two spiral waves do not suppress one another (Fig. 3.14).

$n = 23$

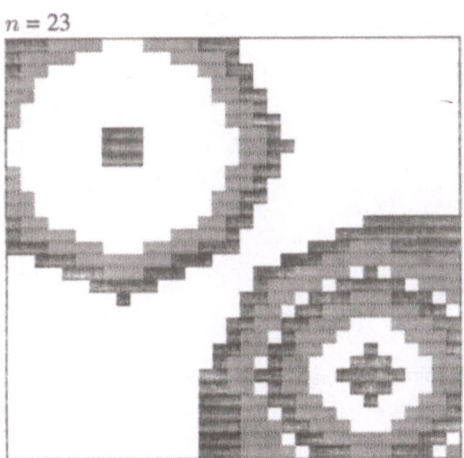

$n = 32$

$n = 90$

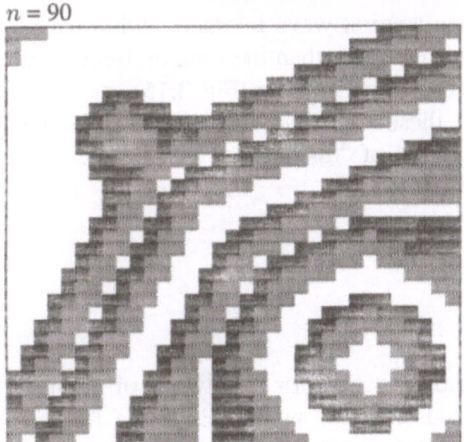

Fig. 3.14. Coexistence of two spiral waves ($\tau_e = 3, \tau_r = 5, g = 0.55, h = 2.5$)

Fig. 3.13. Suppression of a low-frequency pacemaker by a pacemaker with the higher generation frequency ($\tau_e = 3, \tau_r = 4, g = 0.5, h = 3, T_1 = 9, T_2 = 22$)

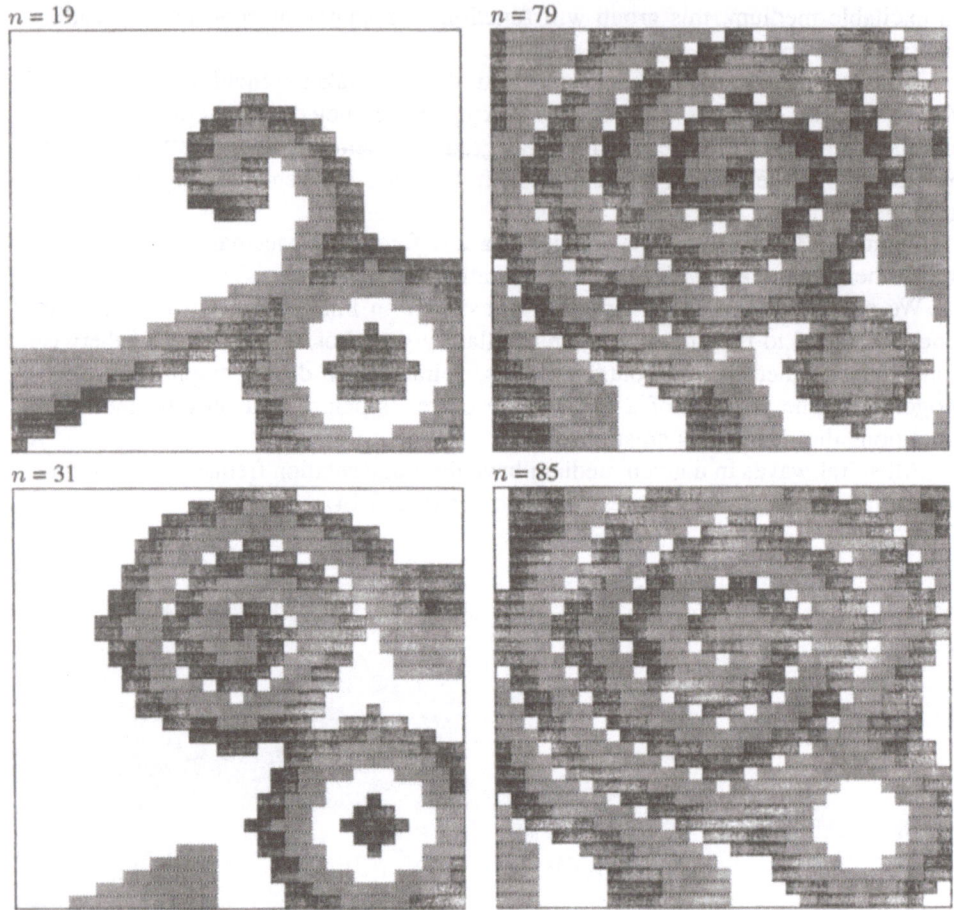

Fig. 3.15. Suppression of a low-frequency pacemaker by a spiral wave ($\tau_e = 4, \tau_r = 5, g = 0.5, h = 3, T = 15$)

Some interesting effects are observed when a spiral wave interacts with a pacemaker. If the generation frequency of a pacemaker is less than the rotation frequency of a spiral wave, the latter completely suppresses a pacemaker (Fig. 3.15).

Under the opposite relationship between these two frequencies a spiral wave finally degenerates into an extra half-wave (a kind of "dislocation") that is steadily pushed out to the periphery of the pacemaker (Fig. 3.16).

3.3 Spiral Waves

Almost all the effects discussed in the previous section for a network of cellular automata have their counterparts in distributed excitable media. Evidence for this is provided both by the results of numeric simulations and by the data from ex-

$n = 11$

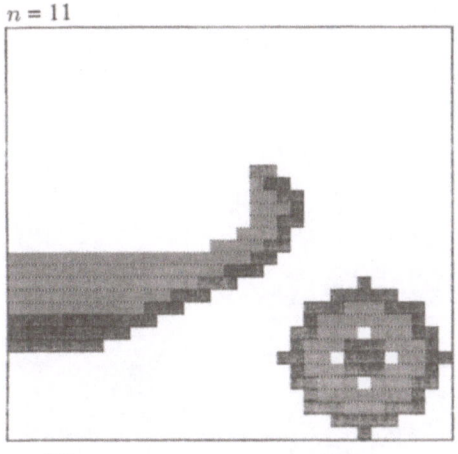

$n = 63$

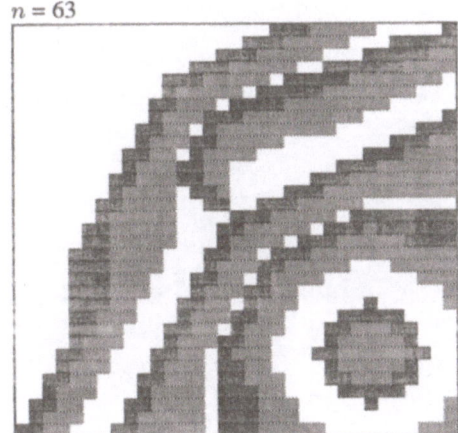

$n = 32$

Fig. 3.16. Ousting of a spiral wave by a high-frequency pacemaker ($\tau_e = 4, \tau_r = 4, g = 0.4, h = 3, T = 6$)

perimental studies. One of the most convenient experimental objects is a chemical excitable medium in the form of a thin unmixed layer of a solution with the Belousov-Zhabotinskii reaction (see [3.1–4]). The advantages of such a system are that there the wave patterns have macroscopic characteristic dimensions (about several millimeters) and their evolution is sufficiently slow (with characteristic time-scales of about a minute). These circumstances considerably simplify observation of various autowave effects. Recently a modification of this reaction, replacing the ferroin as a catalyzer by the ruthenium compound Ru($dipy$)$_3$, was proposed by *Kuhnert* [3.31]. The modified BZ reaction is photosensitive: by changing illumination one can control the effective excitability of the medium.

If we prepare a BZ solution, pour it into a Petri dish and leave it unperturbed for some time, a group of randomly scattered sources of concentrically expanding waves will usually appear (Fig. 3.17). These autowave sources, discovered by *Zhabotinskii* and *Zaikin* [3.32], are called *target patterns*, since they look similar to the targets used in shooting exercises[3]. For a long time it remained unclear whether the target

---

[3] Note that in the Soviet literature another term is common: these patterns are called the "driving centers".

Fig. 3.17. Target patterns in the medium with the Belousov-Zhabotinskii reaction (from [3.32])

patterns are an inherent property of the excitable medium or whether they result from the presence in the solution of certain impurities and gas bubbles which locally change its properties and induce oscillations. Today, however, it is established that target patterns in the BZ reaction are of external origin. In the experiment by *Agladze* and *Krinsky* [3.33] a periodic sequence of plane waves was sent towards a target pattern, with a frequency higher than the frequency of its operation. This led to suppression of a target pattern. However, after the waves ceased to be sent, a target pattern reappeared at exactly the same place.

In contrast to target patterns, spiral waves in the BZ medium, which were discovered by *Zhabotinskii* and *Zaikin* [3.34, 35] and by *Winfree* [3.36], do not require any impurities or inclusions. Usually they are created by breaking the continuous front of an excitation wave. All spiral waves have the same rotation frequency. Therefore, as can be seen from Fig. 3.18, spiral waves coexist, but suppress a target pattern which generates waves at a lower frequency. Multi-armed spiral waves were observed by

Fig. 3.18. Spiral waves and a target pattern in the medium with the Belousov-Zhabotinskii reaction. (From [3.37])

Agladze and *Krinsky* [3.38] in experiments with the BZ reaction (Fig. 3.19; note "touching" of waves in the central region). Recently, *Müller* et al. [3.39–41] performed high-precision computerized experimental studies of spiral waves in the BZ reaction and obtained detailed maps of the reagent distribution.

Spiral waves are the principal type of autonomous wave patterns in homogeneous excitable media. Like vortices in a superconductor or in superfluid helium, they are very stable (this is related to conservation of topological charge, see the discussion below).

Estimation of the ᵣ tion frequency and other properties of spiral waves is an important task of a theory. To start the analysis we consider first the propagation of a travelling pulse in a thin annulus of radius R. If the width of an annulus is very small, the problem is effectively one-dimensional. As pointed out in Sect. 3.1, the solution is the same as the solution for a periodic train of pulses with a spatial period $L = 2\pi R$ on an infinite straight line. The propagation velocity of such a train is determined by its period L, and increases with L, reaching the limit V_0 for $L \to \infty$.

By widening the annulus and allowing its outer radius to tend to infinity, we come to the problem of an excitation wave that rotates around a hole of radius R in an infinite excitable medium. Note that the front of such a wave cannot represent a straight ray rotating with some constant angular velocity ω. If it did so, a segment of the front located at a distance r from the rotation center would have a linear velocity $V = \omega r$ which increases indefinitely for increasing r. But the propagation velocity

Fig. 3.19. Multiarmed spiral waves in the medium with the Belousov-Zhabotinskii reaction. (From [3.38])

of a front cannot exceed V_0. Therefore, the distant parts of a front are retarding and this results in the front curling into a spiral.

Suppose that in polar coordinates a steadily rotating spiral is described by an equation $\varphi = \chi(r) - \omega t$, where ω is the rotation frequency. If we draw a circle of radius r around the center of a spiral, it can be seen immediately (Fig. 3.20a) that the point A, at which the front line crosses this circle, runs along the circle at a velocity $V_{tg} = \omega r$. On the other hand, the motion of the crossing point is caused by

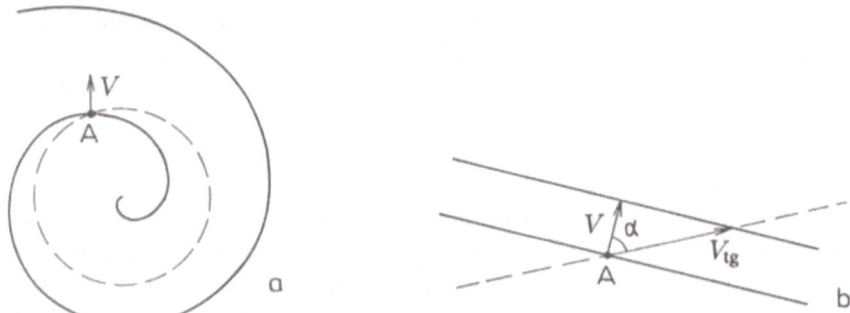

Fig. 3.20a,b. Intersection of a spiral wave front with a given circle (*dashed curve*). The intersection point A moves along the circle with velocity V_{tg}

the propagation of the front in its normal direction at a velocity V. Therefore, we have

$$V_{tg} = V/\cos\alpha \ , \tag{3.3.1}$$

where α is the angle between the front normal and the tangent to the circle at the point A. Simple calculations show that, if the front line is described by $\varphi = \omega t - \chi(r)$ and it passes at a given moment in time through the point A, then

$$\cos\alpha = (1 + r^2\chi_r^2)^{-1/2} \ , \tag{3.3.2}$$

where $\chi_r = d\chi/dr$.

Combining (3.3.1) and (3.3.2), we find

$$V = \omega \left[\chi_r^2 + (1/r)^2\right]^{-1/2} \ . \tag{3.3.3}$$

Suppose now for simplicity that the velocity V of normal propagation is the same for all parts of the front and is equal to the propagation velocity V_0 of a flat excitation wave (i.e. $V = V_0$). Then (3.3.3) represents an equation for the derivative χ_r as a function of a distance r from the spiral wave center at a given value of ω.

The rotation frequency of a spiral wave circulating around a hole of radius R can be estimated by using the following arguments. Since a hole is not permeable for the diffusion flux, the front should be orthogonal to its boundary, which implies $\chi_r(R) = 0$. If we put $r = R$ in (3.3.3) and take this condition into account, we obtain $V_0 = \omega R$. Consequently, the period $T = 2\pi/\omega$ of a spiral wave is equal to the time of a single revolution of a pulse around a circle of radius R, i.e. $T = 2\pi R/V_0$.

Using this result, (3.3.3) can be written as

$$\chi_r^2 + 1/r^2 = 1/R^2 \ . \tag{3.3.4}$$

The function $\chi(r)$ that defines the form of the spiral is found by a simple integration

$$\chi(r) = \int_R^r \sqrt{1/R^2 - 1/r^2} dr \ . \tag{3.3.5}$$

Hence, $\chi(r)$ is monotonously increasing, and therefore in polar coordinates it defines the curve $\varphi = \chi(r)$ which is a spiral. The step h_0 of this spiral can be found from the condition

$$\chi(r + h_0) - \chi(r) = 2\pi \ . \tag{3.3.6}$$

Generally, h depends on distance r from the center. At large distances

$$h_0 \approx 2\pi/\chi_r(r) \ . \tag{3.3.7}$$

Since in this region $\chi_r \approx 1/R$, we find that far from the center the step of a spiral is constant and equal to the perimeter of a hole: $h_0 = 2\pi R$. Note that a spiral with a constant step is called *archimedian*. It can also be noted that the curve, defined in polar coordinates by the equation $\varphi = \chi(r)$ with the function $\chi(r)$ given by (3.3.5), represents an involute of a circle of radius R.

Thus, we have found that a spiral wave can rotate steadily around a hole in an excitable medium, and we have determined in the simplest approximation its form and its rotation frequency.

Our analysis was approximate in two respects. First, we neglected the dependence of the propagation velocity of waves on the interval between them. This deficiency can easily be corrected. Since the angular velocity of rotation is ω, the entire pattern is completely repeated at time moments separated by $T = 2\pi/\omega$. This implies that the interval between the passage of any two consecutive waves through a given point of the medium is also equal to T. Therefore, instead of V_0 we should substitute the velocity $V(T)$ for propagation of a pulse train with a temporal period T. The period T itself can then be found by solving the equation

$$V(T)T = 2\pi R .\tag{3.3.8}$$

The second circumstance is more subtle, and it cannot be so easily taken into account. The above arguments neglected the dependence of the propagation velocity of a front on its local curvature. Actually, the parts of a front located at different distances from the center of a spiral wave have different curvatures, and therefore their normal propagation velocities will be different too.

The local curvature K of the curve defined in polar coordinates by an equation $\varphi = \chi(r)$ is (see [3.42])

$$K = (r\chi_{rr} + 2\chi_r + r^2\chi_r^3)(1 + r^2\chi_r^2)^{-3/2} .\tag{3.3.9}$$

Using (3.3.4) we find that the second derivative $\chi_{rr} = \partial\chi_r/\partial r$ is

$$\chi_{rr} = (r^2\chi_r)^{-1} .\tag{3.3.10}$$

Since the front comes orthogonally to a hole, $\chi_r \to 0$ at $r \to R$. But then (according to (3.3.10)) $\chi_{rr} \to \infty$ as $r \to R$, and therefore the front curvature of a spiral wave diverges (cf.(3.3.9)) at the hole boundary. However, we are justified in neglecting the dependence of the propagation velocity V on the front curvature K only provided that this curvature is sufficiently small. We see that near the hole boundary this is not so.

It will be shown in the next section that the corrected solution, taking into account such dependence, differs from the solution found above only within a narrow layer of width about D^2R/V_0^2 near the hole boundary; outside this narrow layer the spiral has the form of a circle involute. For the corrected solution, the front curvature does not diverge at the hole boundary, but approaches a finite value of about $(V_0/DR^2)^{1/3}$.

Note that the problem of spiral wave circulation around a hole was first analyzed under the assumption of a constant propagation velocity by *Wiener* and *Rosenblueth* [3.20] who found that a spiral wave front represents an involute of the hole.

The examples of Sect. 3.2 show that spiral waves can also be observed in a homogeneous medium without any holes.

The tip of a spiral wave in such a medium rotates around a circular region which can be called the *core* of this wave. Inside the core, the medium remains quiescent, despite the fact that this region does not differ in its properties from other parts of the same excitable medium. The core looks like an effective hole. Now the principal problem is: what prevents excitation from entering into the core region?

It was noted above that for every medium there is a minimal possible period L_{min} of a propagating wave train. Taking into account similarity between a periodic wave train on a straight line and a solitary pulse on a circle, we can conclude that stable pulse propagation is impossible on circles with radii less than $R_{min} = L_{min}/2\pi$.

Let us take a medium with a hole, around which a spiral wave is rotating, and begin to diminish its radius R. What will happen when the hole radius becomes smaller than R_{min}?

An excitation wave cannot circulate around such a small hole with its front sticking to the border. Therefore the spiral wave becomes separated from the hole and its free tip appears. The minimal possible radius of a circle along which such a tip can steadily circulate is R_{min}. This minimal radius can be taken as a rough estimate of the size of a spiral wave core. In this approximation, spiral waves have rotation frequency $\omega = \omega_{max}$. The spirals are curled in a maximally tight way since their step far from the center is $h_0 = L_{min}$.

This simple estimate of the rotation frequency of spiral waves and of their core radius agrees as to order of magnitude with the simulation data (see e. g. [3.19, 43]) in the case of sufficiently high excitability; it predicts a rotation period of a spiral wave that is about two times smaller than the actual one. The origin of the remaining discrepancy is clear. When deriving this estimate, we neglected the effects of the front curvature of a spiral wave, i. e. its influence on the propagation velocity. In reality, those parts of a spiral wave which are closer to the core have higher curvatures, and therefore they move slower than a flat front or a solitary pulse. As a result, the rotation frequency of a spiral wave is decreased in comparison with a simple estimate ω_{max}. Better estimates for the rotation frequency were obtained in [3.13, 44].

However, spiral waves with a small core (of radius about R_{min}) and a small rotation period (about T_{min}) are observed only in media with high enough excitability. When the excitability of a medium is lowered, another circulation regime is gradually established which is characterized by a large period ($T \gg T_{min}$) and a large core radius ($R \gg R_{min}$) of spiral waves. The description of such spiral waves and estimation of their principal properties requires a special analysis, carried out in the next section.

3.4 Front Kinematics

Since excitable media are nonlinear, propagating waves interact. However, in excitable media this interaction is local. If one wave follows after another, it does not feel the presence of a previous wave while the distance between them is much larger than the characteristic width of an excitation pulse wave, a length of about L_{min}. In this section we restrict ourselves to the analysis of situations when every wave is separated from all others by distances much larger than L_{min}. Then we can neglect the finite width of a single wave and assume that it is completely specified by an oriented curve of its wavefront.

Since the states of a medium before and after the passage of an excitation pulse are the same, a wavefront can be broken off somewhere inside the medium, so that a tip is created (cf. Sect. 3.2). Besides moving in the normal direction, a wavefront can sprout or contract at its tip. The velocity G of such tangent motion depends on the curvature K_0 of a wavefront close to the tip. Suppose that G_0 is the speed of the tip sprouting of a flat wave. Then, for sufficiently small curvatures K_0, the speed of sprouting G for a curved front should have a small correction linear in K_0, i.e. $G = G_0 - \gamma K_0$ (*Brazhnik* et al. [3.45] showed that this can be proved using perturbation theory). We consider below only those excitable media where the speed of sprouting decreases with increasing curvature (i.e. $\gamma > 0$); otherwise a medium has no stable spiral wave solutions.

Cellular automata simulations in Sect. 3.2 reveal that, by changing the excitability of a medium, one can control the tip motion, forcing the wave to elongate or to contract. The same effect is found in realistic reaction-diffusion systems described by partial differential equations. Figure 3.21 shows the successive evolution of a broken excitation wave found by a computer simulation of a certain reaction-diffusion model at four different values of a parameter that specifies the medium excitability.

Thus, by changing excitability, we can vary the speed of sprouting G_0, and even reverse its sign. Note that when G_0 is negative, a broken wave always contracts at its tip and finally disappears. Below we assume that G_0 is positive.

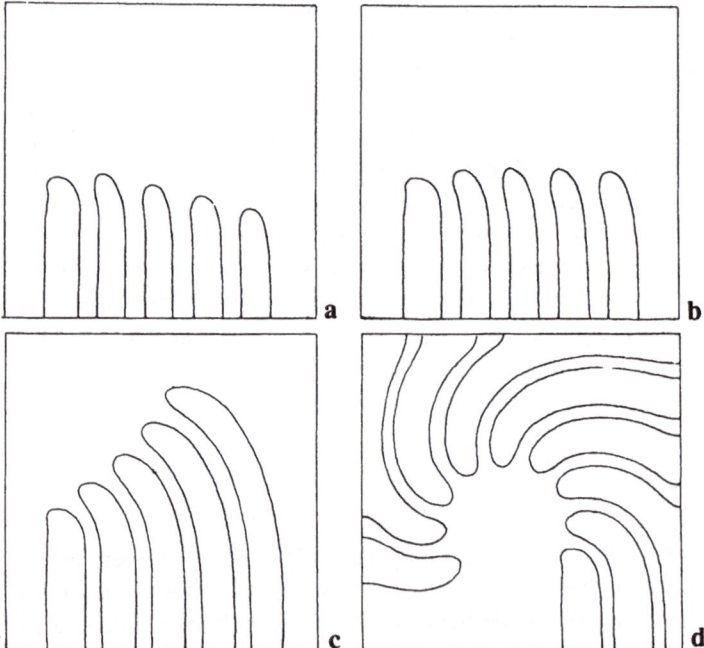

Fig. 3.21a–d. Evolution of a broken excitation wave for a two-component reaction-diffusion model (7.36) in the monograph by *Zykov* [3.19]. Lines of constant activator concentration E are shown at subsequent time moments for the four different excitabilities of the medium, controlled by the parameter ε. (a) poor excitability ($\varepsilon = 0.4$), (b) marginal excitability ($\varepsilon = 0.388$), (c) supercritical excitability ($\varepsilon = 0.35$), (d) high excitability ($\varepsilon = 0.3$). (From [3.46])

To simplify the consideration, we assume also that the linear dependence of G on K is holds up to the value $K_{cr} = G_0/\gamma$ of the front curvature near the tip at which the speed of sprouting becomes equal to zero. It is also supposed that $DK_{cr} \ll V_0$. These two assumptions imply the existence of a small parameter

$$\beta = \sqrt{DK_{cr}/V_0} . \tag{3.4.1}$$

The assumptions are satisfied when excitability of a medium is sufficient for the existence of spiral waves, but is still not too high. Many results can be generalized to a case when these two assumptions are violated, but then calculations become more tedious.

Kinematics of wavefronts in two-dimensional excitable media can be formulated as follows[4]. Any wave is specified then by indication of the curve of its front. Most conveniently this can be done using a *natural equation* $K = K(l)$ of the curve which gives the dependence of the front curvature K on the length l of the curve arc, measured from the end point (i.e. from the wave tip). Note that any natural equation specifies the curve up to its position on a plane. If the form of a curve changes in time, $K = K(l, t)$.

The evolution of a broken wavefront is determined by the two conditions:
a) At each time moment any small segment of a front moves in its normal direction with the velocity

$$V = V_0 - DK . \tag{3.4.2}$$

b) The front contracts ($G < 0$) or smoothly sprouts in the tangent direction at its tip with the velocity

$$G = \gamma(K_{cr} - K_0) , \tag{3.4.3}$$

where $K_0 = \lim_{l \to 0} K(l)$.

It turns out that these two conditions are sufficient to find the equation for $K(l, t)$. A rigorous derivation can be given using the methods of differential geometry (see [3.51]). Below we choose a simpler derivation given by *Davydov* and *Mikhailov* [3.51].

Let us consider a wavefront at some moment t. Suppose that its curvature at a point a is K_a (Fig. 3.22). After a small time interval dt, a small segment in the vicinity of this point will be moved into a segment in the vicinity of a point b, and its curvature will become equal to K_b. Let us find the relationship between K_a and K_b. In order to do this, it is convenient to go to a polar coordinate system with its pole in the curvature center of a small front segment including point a. In this system we have

$$\varrho_b = \varrho_a + V \, dt , \tag{3.4.4}$$

[4] The kinematical approach was first used in 1951 by *Burton* et al. [3.47] to describe the phenomenon of spiral growth of crystals around screw dislocations. It was applied in the theory of wave patterns in excitable media by *Zykov* [3.19, 48, 49] who did not initially take into account the effect of the wavefront sprouting.

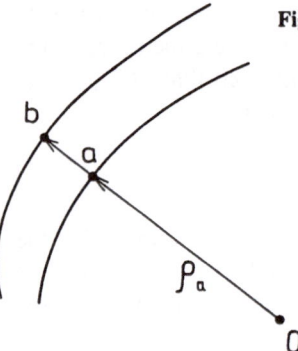

Fig. 3.22. Front positions at two subsequent time moments

where $\varrho_a = 1/K_a$ is the curvature radius of the front at the point a, and ϱ_b is the polar radius of the point b.

The curvature of a curve, given in polar coordinates by equation $\varrho = \varrho(\varphi)$, can be calculated as (see [3.42])

$$K = \frac{\varrho^2 - 2\varrho' - \varrho\varrho''}{\left(\varrho^2 + \varrho'^2\right)^{3/2}} .$$
(3.4.5)

Substituting (3.4.4) into (3.4.5) and taking into account that $\partial \varrho_a / \partial \varphi = 0$, we find in the first order in dt:

$$K_a = K_b - V K_a^2 dt - K_a^2 (d^2 V/d\varphi^2) dt .$$
(3.4.6)

Replacing further the derivatives with respect to the polar angle φ by the derivatives with respect to the length l of the arc $(dl = \varrho_a d\varphi)$, we obtain for $dK = K_a - K_b$ that

$$dK = - \left[K_a{}^2 V + \left(d^2 V/dl^2\right)\right] dt .$$
(3.4.7)

On the other hand, $dK = (\partial K/\partial l)dl + (\partial K/\partial t)dt$. The increment dl of the arc length in time dt is given by

$$dl = \left(\int_0^l KV d\xi\right) dt + G dt .$$
(3.4.8)

The first term in (3.4.8) describes the elongation of the curve in the process of its normal expansion; the second term takes into account changes of the arc length due to sprouting or contraction (for $G < 0$) of the curve at its free end which is taken as a reference point for l.

If we combine these two alternative expressions for dK, we come to the equation

$$\frac{\partial K}{\partial t} + \left(\int_0^l KV d\xi + G\right) \frac{\partial K}{\partial l} = -K^2 V - \frac{\partial^2 V}{\partial l^2} ,$$
(3.4.9)

where the velocity G of sprouting is determined by the curvature $K(0,t) = \lim_{l\to 0} K(l,t)$ at the free end, and the velocity V of the normal front motion is determined by the local curvature K of the front, i.e. $V = V(K(l,t))$. Thus we obtain a closed equation for $K(l,t)$ that governs the evolution of a broken wavefront[5]. Specifically, if G and V are given by (3.4.2) and (3.4.3), we find

$$\frac{\partial K}{\partial t} + \left[\int_0^l KV \, d\xi + \gamma(K_{cr} - K(0,t)) \right] \frac{\partial K}{\partial l} = -K^2 V_0 + D \frac{\partial^2 K}{\partial l^2} \, . \qquad (3.4.10)$$

This equation is derived under the assumption that all curvatures are sufficiently small to satisfy the condition $DK/V_0 \ll 1$. Therefore, we have replaced V by V_0 everywhere in (3.4.10), except for the last term. Although it is proportional to a small parameter β (see (3.4.1)), this term involves a higher derivative and we should be particularly careful in dealing with it. Below it is shown that such a term does play an important role in a spiral wave solution.

Equation (3.4.10) has a trivial stationary solution

$$K(l,t) = 0 \quad \text{for} \quad l \geq 0 \, ,$$

that corresponds to a flat broken front which propagates with velocity V_0. However, this solution is absolutely unstable with respect to small perturbations. Numeric simulations of (3.4.10) reveal (Fig. 3.23) that, independent of the form of an initial perturbation, a flat halfwave always evolves to a steadily rotating spiral wave.

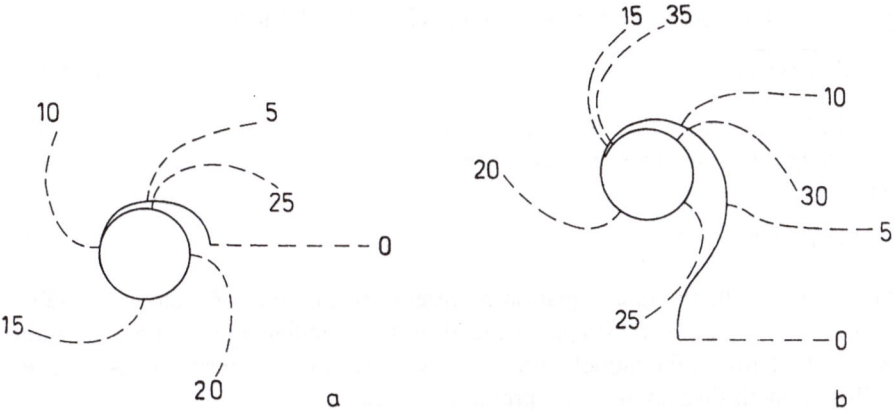

Fig. 3.23. Evolution of initially flat broken wave with a small positive (a) or negative (b) curvature perturbation near its end. Subsequent front positions (dashed lines) are shown; the solid line indicates the trajectory of the tip motion. (From [3.46])

---

[5] This fundamental equation of kinematics was first derived by *Zykov* [3.52] (see also [3.19] where it is decomposed into two related equations). Similar equations were derived by *Brower* et al. [3.53] in the theory of interface evolution; however their models did not include the effects of tangential growth. *Meron* and *Pelcé* [3.54] used this equation to describe formation of spiral waves, assuming that the speed of tangent sprouting is constant.

By using (3.4.10), we can find the basic properties of steadily rotating spiral waves.

For a spiral wave that rotates at a fixed angular velocity, the form of a front is constant; only its position on the plane is changed in time. Therefore, K does not depend on t. Moreover, stable steady circulation is possible only if the front neither grows nor contracts at its tip. This implies that the front curvature at the end point should be critical, i.e. $K(0) = K_{cr}$. The circle around which the tip moves is a core of the spiral wave; note that the front is orthogonal to the core.

Equation (3.4.10) also describes the circulation of a spiral wave around a hole in an excitable medium. If the hole boundary is not permeable for the diffusion flux, the front is orthogonal to it. The form of the wavefront is then determined by a stationary limit of (3.4.10) in the absence of the term that describes sprouting of the tip (there is no free tip in this case since the front touches a hole).

For a steady circulation (when $\partial K/\partial t = 0$) equation (3.4.10) can be integrated once, which gives

$$K \int_0^l K V_0 d\xi - D\frac{\partial K}{\partial l} = \omega .\tag{3.4.11}$$

The integration constant ω is simply the rotation frequency of the spiral wave. This can be easily recognized, if we note that the front meets orthogonally with the core circle (or with the hole). Hence, we see that $dV/dl = -D(dK/dl)$ at $l = 0$ is the rotation frequency of a spiral wave front.

First we consider what is found in a model where V does not depend on the curvature, so that $D = 0$ and the second term on the left side of (3.4.11) vanishes. Then equation (3.4.11) can be easily solved analytically. Its solution is

$$K = \sqrt{\omega/2lV_0} .\tag{3.4.12}$$

Suppose that a hole of radius R is present in the medium. Then, the rotation frequency of a spiral wave that circulates around this hole is equal to $\omega = V_0/R$ and (3.4.12) gives

$$K = (2Rl)^{-1/2} .\tag{3.4.13}$$

A curve defined by this natural equation represents an involute of a circle of radius R. Far from a hole (at $l \gg R$) it approximates an archimedian spiral with a constant step $h_0 = 2\pi R$. Thus, in the model with $V = 0$ we reproduce the results of N. Wiener and A. Rosenblueth discussed in the previous section.

Note that at $l \to 0$ the solution (3.4.13) has $K \to \infty$. This implies that, if the coefficient D in the dependence $V = V_0 - DK$ is small but nonvanishing, the dependence of the propagation velocity on the curvature cannot be neglected for sufficiently small l, close to the boundary of a core or a hole.

Now we turn to the problem of spiral waves in a homogeneous medium (in absence of any holes) taking into account the dependence of the propagation velocity of a front on its local curvature. The rotation frequency $\omega = \omega_0$ of a spiral wave, that enters as an unknown parameter into (3.4.11), is determined in this case from the following arguments.

Far from the center of a spiral, the front curvature should tend to zero, i.e. $K \to 0$ as $l \to \infty$. In effect, this condition already defines in a unique way a certain phase trajectory for the differential equation (3.4.11) at a given value of ω. Thereby, the value of the curvature $K(0)$ at $l = 0$, i.e. near the tip, is uniquely determined. Demanding now that $K(0) = K_{cr}$, we obtain an equation which allows us to find ω_0.

For sufficiently large l, we can neglect the term with the derivative on the right side of (3.4.11). In this region the solution is given by (3.4.12). On the other hand, it follows from (3.4.11) that at $l = 0$ we should have $dK/dl = -\omega/D$. Therefore, for small l the natural equation of our curve is

$$K \approx K_{cr} - (\omega/D)l . \tag{3.4.14}$$

In effect, (3.4.12) and (3.4.14) represent the outer and the inner approximations for the same phase trajectory, and therefore they should match one another. Let us require that at some point $l = l_0$ the values of both functions, as well as their derivatives, coincide. These conditions give two equations:

$$K_{cr} - (\omega/D)l_0 = (2V_0 l_0/\omega)^{-1/2} ,$$
$$\omega/D = (1/2)(2V_0/\omega)^{-1/2} l_0^{-3/2} . \tag{3.4.15}$$

Their solution is

$$\omega_0 = (2/3)^{3/2} \sqrt{DV_0} K_{cr}^{3/2} , \tag{3.4.16a}$$
$$l_0 = (1/2)(3/2)^{1/2} \sqrt{D/K_{cr} V_0} . \tag{3.4.16b}$$

A more accurate estimate for the rotation frequency ω_0 of spiral waves can be obtained by scaling arguments.

Introducing new units for its variables, equation (3.4.11) can be written in a universal form, not involving any parameters of a particular medium. It defines a universal dimensionless rotation frequency Λ that is independent of the properties of an excitable medium. By numeric integration one can find the value of this universal parameter. Afterwards, the dimensional rotation frequency is easily recovered. Such an analysis yields

$$\omega_0 = \Lambda \sqrt{DV_0} K_{cr}^{3/2} , \quad \Lambda \approx 0.69 \ldots , \tag{3.4.17}$$

which differs from (3.4.16a) only in the more accurate value of the numeric factor.

This result was first found by *Zykov* [Ref. 3.19, Sect. 3.6] who obtained also further corrections in powers of β. However, in [3.19, 3.48] *Zykov* treats K_{cr} as the critical curvature K^* for propagation of the solid front (see Fig. 3.5). Actually, K_{cr} is smaller than K^*, but in some cases their values can be very close. The corrected expression was derived in [3.51, 3.55].

Up to the terms of order DK_{cr}/V_0, the core radius R_0 of a spiral wave can be estimated from the condition $\omega_0 R_0 = V_0$ as

$$R_0 = \Lambda^{-1} \sqrt{V_0/D} K_{cr}^{-3/2} . \tag{3.4.18}$$

Now we can discuss the results. We see that as $K_{cr} \to 0$ the rotation frequency ω_0 of a spiral wave diminishes and tends to zero; at the same time the core radius R_0

of the spiral wave increases indefinitely. Everywhere except in a narrow boundary layer with a width of about l_0 near the core border, the wavefront of the spiral represents an involute of a circle of radius R_0; its form is given by (3.4.13) with $R = R_0$. The width l_0 of the boundary layer is small compared with the core radius, i. e. $l_0/R_0 \sim (DK_{cr}/V_0) \ll 1$.

In a similar way we can find a solution for a spiral wave rotating around a hole. It turns out that, at the border of a hole of radius R, a wavefront has the curvature

$$K(0) = \Lambda^{-2/3}\left(V_0/DR^2\right)^{1/3}, \qquad (3.4.19)$$

provided that $R \gg D/V_0$. Outside a narrow boundary layer with a width of about D^2R/V_0^2, the spiral has the form of an involute of the hole. With an accuracy to the terms of the first order in $(D/V_0R)^{2/3}$, the rotation frequency ω is determined by the condition $\omega R = V_0$.

Note that the tip of a spiral wave that moves while sticking to an impermeable boundary of the hole cannot be considered a free one. Since there is no diffusion flux in the direction along the front at its end, the conditions of front propagation are the same as those for the solid, unbroken front. At the hole boundary the front curvature can therefore exceed the critical value K_{cr} for the free end of a wavefront, but it should remain less than the critical curvature K^* for the propagating unbroken front (see (3.2.18) and Fig. 3.5). This explains the hysterisis effect observed by *Pertsov* et al. [3.43] in computer simulations.

In a certain interval of the hole radii ($R^* < R < R_0$), two circulation regimes with different rotation frequencies coexist. In the first regime, the end of a wavefront moves along the boundary of a hole, sticking to it. In the second regime, the wavefront has a free end that moves along a large core of radius R_0 and does not feel presence of a hole located in the center. If a large hole of radius $R > R_0$ had been created in the medium at an initial time moment and if later we gradually

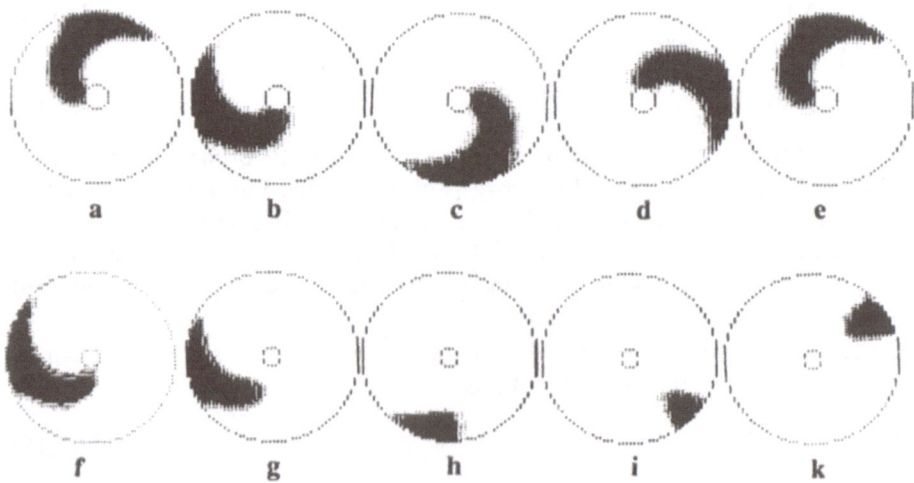

Fig. 3.24. Steady rotation (a-e) of the wave around a hole with the larger radius. Transition (f-k) to slow rotation with a separated wave tip after the hole radius was decreased (from [3.43])

diminish its size, the tip of the spiral wave will remain stuck to the hole until the value $R = R^*$ is reached. After that, the tip will separate from the hole and the rotation frequency will abruptly decrease, reaching the value (3.4.17) of the rotation frequency of spiral waves in homogeneous media (Fig. 3.24). The latter value of the rotation frequency will be maintained for $R < R^*$. If we go in the reverse direction, starting from a vanishingly small hole and then slowly inflating it, the regime with a free tip of a spiral wave is maintained until $R = R_0$; the tip of a spiral wave becomes stuck to the hole border only when $R > R_0$.

In the simple kinematical model it is assumed that the normal propagation velocity V depends only on the local curvature of a given segment of the wavefront, i.e. $V = V(K)$. Generally, this velocity V can also depend on the distance from the end point (i.e. on the arc length l) so that we have

$$V = V_0(l) - DK .\tag{3.4.20}$$

This modification can be easily incorporated to the theory.

Note that such a dependence can hold only for short distances from the free end. Let l^* be the characteristic distance to which it persists. We can analyze two possible situations (see [3.50]).

Suppose $l^* \ll l_0$. Then only the form of the solution within the boundary layer changes. By an appropriate matching, we obtain

$$\omega_0 = \Lambda\sqrt{DV_0}\left[K_{cr} + D^{-1}\left(V_0^\infty - V_0(0)\right)\right]^{3/2} ,\tag{3.4.21}$$

i.e., the rotation frequency of the spiral wave increases.

For $l^* \gg l_0$, the form of the solution in the outer region is changed:

$$K = \sqrt{(\omega_0/2)\int_0^l V_0(\xi)d\xi} .\tag{3.4.22}$$

After the matching we find that the rotation frequency is still given by (3.4.17) but with V_0 replaced by $V_0(0)$.

Note that all results found in this section are applicable only in the case when the core radius R_0 of a spiral wave is large as compared with $R_{min} = L_{min}/2\pi$ (see Sect. 3.1), i.e. when the core perimeter is much larger than the width of a single pulse (the front does not run into its own tail).

Keener and *Tyson* [3.56–59] generalized the kinematic theory to include the dependence of the propagation velocity on the residual inhibitor concentration left by a previous pulse. This allowed them to consider spiral waves with smaller cores (which are observed, for instance, under standard conditions in the BZ chemical solution). However, their approach did not take wavefront sprouting into account and therefore they were able to consider only the effects of steady circulation of spiral waves.

3.5 Resonance, Drift, and Meandering of Spiral Waves

The fundamental equation of front kinematics (3.4.10) can be used to study various time-dependent wave phenomena in excitable media.

We consider first the process of establishing a steady circulation of spiral waves in homogeneous media. Suppose that a small perturbation is introduced into the form of a spiral wave, localized at a distance l from the free end of the wavefront. As follows from (3.4.10), this perturbation will drift away from the center to the periphery, simultaneously spreading and fading in a diffusional manner, with an effective "diffusion" constant D. According to (3.4.10), near the tip (where $K \approx K_{cr}$) the velocity of the drift can be approximately estimated as $K_{cr}V_0 l$.

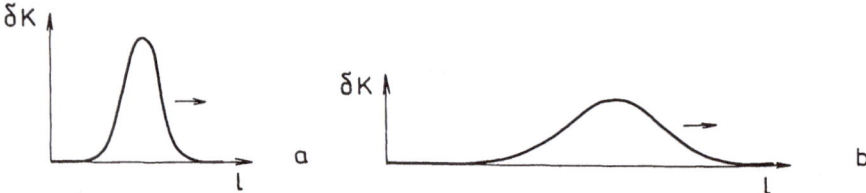

Fig. 3.25a,b. An initial perturbation (a) and its form at a later time (b)

Hence, this perturbation moves away from the tip at a speed of about $K_{cr}V_0 l$ but at the same time its width increases as $(Dt)^{1/2}$ (see Fig. 3.25). What is the distance at which such a perturbation should be localized initially if, despite the drift, it reaches the end point of the front? The latter happens if the width $L_s \approx (Dt)^{1/2}$ of a spreading perturbation exceeds (during some time interval) the distance $L_c \approx l + K_{cr}V_0 lt$ from its center to the end point of a wavefront. Comparing these two expressions, we find that L_s can exceed L_c only while the initial distance l is less than $(D/K_{cr}V_0)^{1/2}$. But, according to (3.4.16b), this is precisely the length l_0 of a boundary layer near the core!

Consequently, the tip motion can be influenced only by those front perturbations that are created within a narrow boundary layer, at a distance not larger than l_0 from the end point. These perturbations of the front form are damped within a time

$$\tau_D \sim l_0^2/D \sim (K_{cr}V_0)^{-1} . \tag{3.5.1}$$

Note that the characteristic damping time τ_D for perturbations of the wavefront form that can influence the tip motion is always smaller than the rotation period of a spiral wave. Indeed, it follows from (3.4.17) and (3.5.1) that

$$\omega_0\tau_D \sim \sqrt{DK_{cr}/V_0} \ll 1 . \tag{3.5.2}$$

There is also another characteristic time τ_G related to the effects of front sprouting or contraction.

If the coefficient γ in the expression (3.4.3) for the speed of sprouting were equal to zero, K_{cr} would not enter at all into the fundamental equation of kinematics

(3.4.10). In this case, many spiral wave solutions exist, differing by the value $K(0)$ of the front curvature at a free end. Every such solution describes a certain spiral wave that steadily circulates with a frequency ω given by the variant of (3.4.17) obtained by replacing K_{cr} by $K(0)$. Then, by locally perturbing the form of a wavefront within the boundary layer, we can initiate a transition to a new circulation regime with the value of the front curvature at a free end differing from the previous one by a small correction $\delta K(0)$.

When the coefficient γ is nonvanishing and positive, all deviations of the curvature $K(0)$ from K_{cr} are damped. However, when γ is sufficiently small, the characteristic time-scale τ_G of such damping is large and a *quasistationary regime* with $\tau_G \gg \tau_D$ is realized. In this regime, which was investigated by *Brazhnik* et al. [3.60], the form of a wavefront near the core adjusts adiabatically to the curvature $K(0, t)$ at the free end. In its turn, this curvature slowly varies because of sprouting or contraction according to the equation

$$\left. \frac{\partial K}{\partial t} \right|_{l=0} = -G \left. \frac{\partial K}{\partial l} \right|_{l=0} . \tag{3.5.3}$$

The derivative $(\partial K/\partial l)$ at $l = 0$ is determined by

$$\left. \frac{\partial V}{\partial l} \right|_{l=0} = -D \left. \frac{\partial K}{\partial l} \right|_{l=0} = \omega , \tag{3.5.4}$$

where ω is given by (3.4.17) with replacement of K_{cr} by $K(0, t)$.

Combining all these remarks, we find that under the conditions of a quasistationary regime the curvature $K_0 = K(0, t)$ at a free end obeys the equation

$$\frac{dK_0}{dt} = -\Lambda \gamma \sqrt{V_0/D} K_0^{3/2} (K_0 - K_{cr}) . \tag{3.5.5}$$

For small deviations $\delta K_0 = K_0 - K_{cr}$, Eq. (3.5.5) can be linearized so that it takes the form

$$(d/dt)\delta K_0 = -\delta K_0/\tau_G , \tag{3.5.6}$$

where the characteristic relaxation time τ_G is

$$\tau_G = D/\gamma\omega_0 . \tag{3.5.7}$$

Note that $\omega_0 \tau_G = D/\gamma$, and therefore the quasistationarity condition $\tau_D \ll \tau_G$ reads

$$\gamma/D \ll \sqrt{V_0/DK_{cr}} . \tag{3.5.8}$$

Estimates for some known examples of excitable media show that γ usually varies from zero to about D. Therefore, since $(DK_{cr}/V_0)^{1/2} \ll 1$ is assumed, condition (3.5.8) is almost always satisfied and the quasistationary approximation is applicable.

If $\gamma \ll D$, we have $\omega_0 \tau_G \gg 1$. Then the process of settling into steady circulation of a spiral wave, which includes relaxation of the front curvature at the free end to

K_{cr}, goes on for many rotations of the wave. Spiral waves in such excitable media are *highly inertial*. On the other hand, when the opposite condition $\gamma \gg D$ holds, steady circulation is established in a time which is small compared to the period of a single rotation. In this case spiral waves have *low* inertiality.

As mentioned in Sect. 3.4, a natural equation $K = K(l,t)$ defines only the form of a curve, but not its position on a plane. To specify the evolution of a wavefront, we should supplement this equation by equations which determine the orientation of this curve and indicate the position of its initial point.

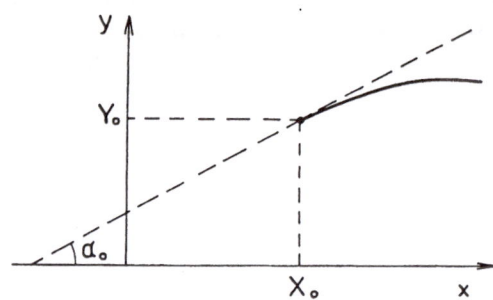

Fig. 3.26. Indication of X_0, Y_0 and α_0 fixes the position of the end point and the orientation of the front curve

Let us introduce a cartesian coordinate system (x, y) on a plane. Suppose that $X_0(t)$ and $Y_0(t)$ are the coordinates of the tip of a wave at a time moment t, and $\alpha_0(t)$ is the angle between the tangent to the front at the end point and the x-axis (see Fig. 3.26). One can easily recognize that X_0, Y_0, and α_0 are sufficient to fix the position of a front in the plane.

Since the end point moves simultaneously in the normal direction at a velocity $V(0) = V_0 - DK_0$ and in the tangent direction at a velocity G (because of front sprouting or contraction), its cartesian coordinates change in time according to the equations

$$
\begin{aligned}
\dot{X}_0 &= -V(0)\sin\alpha_0 - G\cos\alpha_0 \ , \\
\dot{Y}_0 &= V(0)\cos\alpha_0 - G\sin\alpha_0 \ .
\end{aligned}
\tag{3.5.9}
$$

Furthermore, the evolution of the angle α_0 that specifies the direction of a front tangent at the end point obeys an equation

$$
\frac{\partial \alpha_0}{\partial t} = \frac{\partial V}{\partial l}\Big|_{l=0} + GK_0
\tag{3.5.10}
$$

The first term here takes into account rotation (note that $\partial V / \partial l$ at $l = 0$ is the angular velocity of the end point), while the second term describes changes of α due to the tangent sprouting of the front. Since $V = V_0 - DK$, we obtain

$$
\frac{\partial \alpha_0}{\partial t} = -D\frac{\partial K}{\partial l}\Big|_{l=0} + GK_0
\tag{3.5.11}
$$

Together with the fundamental equation of kinematics (3.4.10), equations (3.5.9) and (3.5.11) completely determine the motion of a wave with a free tip.

The law of the tip motion is especially simple in a quasistationary approximation, when we can distinguish the effects of sprouting from the effects of establishing the steady form of a wavefront near the core. In this case $\partial K/\partial l \approx -\omega/D$ at $l = 0$, where ω is given by (3.4.17) with the replacement of K_{cr} by K_0. Hence, (3.5.11) reads as

$$\dot{\alpha}_0 = \Lambda\sqrt{V_0 D}K_0^{3/2} + \gamma K_0(K_{cr} - K_0) . \tag{3.5.12}$$

Therefore, when condition (3.5.8) holds, the trajectory of motion of the tip of a spiral wave can be found by solving four ordinary differential equations (3.5.5), (3.5.9), and (3.5.12).

By varying the parameters of a medium, we can change its excitability and, therefore, the critical curvature K_{cr}. Suppose that this quantity changes periodically in time as

$$K_{cr}(t) = K_{cr} + K_1 \cos(\omega_1 t + \varphi) , \tag{3.5.13}$$

where $K_1 \ll K_{cr}$ and the modulation frequency ω_1 is close to the rotation frequency ω_0 of a spiral wave. Then, as shown by *Brazhnik* et al. [3.55, 60], a *resonance effect* should be observed.

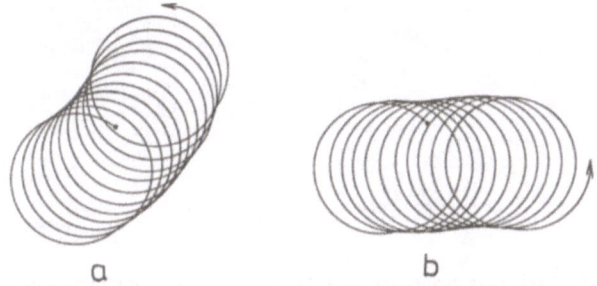

a   b

Fig. 3.27a,b. Resonance of a spiral wave under periodic modulation of the medium excitability. The trajectory of the free end point is shown. The spiral wave frequency is $\omega = 0.3186$, and the parameter K_{cr} varies periodically with frequency (a) $\omega_1 = 0.315$ and (b) $\omega_1 = 0.32$. (From [3.46, 60])

In the presence of a periodic modulation of K_{cr}, the rotation center of a spiral wave does not rest but performs a circular motion. This effect can be revealed (see Fig. 3.27) by a direct numeric integration of equations (3.5.5), (3.5.9), and (3.5.12) with $K_{cr}(t)$ given by (3.5.13). When $K_1 \ll K_{cr}$, the approximate analytical theory of the resonance effect is also possible. In this case, variations of K_0 are small and the linearized version of these equations can be used. Straightforward, but somewhat tedious, calculations yield the following results.

The radius R_{res} of a circle along which the center of a spiral wave moves is

$$R_{res} = \frac{(3/4)V_0(K_1/K_{cr})}{|\omega_1 - \omega_0|\sqrt{1 + (D/\gamma)^2}} . \tag{3.5.14}$$

Note that R_{res} grows when ω_1 is approaching ω_0. The velocity of motion of a spiral wave center is

$$V_{\text{res}} = \frac{(3/4)V_0(K_1/K_{\text{cr}})}{\sqrt{1+(D/\gamma)^2}} \qquad (3.5.15)$$

and is proportional to the modulation amplitude K_1.

When the two frequencies coincide (i. e. under the condition $\omega_1 = \omega_0$ of complete resonance), the spiral wave center moves at a constant velocity (3.5.15) along a straight line. The direction of the center motion is then determined by the initial modulation phase φ and by the direction of rotation of this spiral wave (clockwise or counterclockwise).

Resonance of spiral waves was observed by *Agladze* et al. [3.61] in the experiment using a photosensitive modification of the Belousov-Zhabotinskii reaction. In the initial set-up, one or two spiral waves were created in the medium. Then, after measuring the period of steady circulation of a spiral wave, the liquid solution was periodically illuminated by an intense uniform light source. The evolving wave pattern was recorded after equal time intervals on the background of two fixed reference lines.

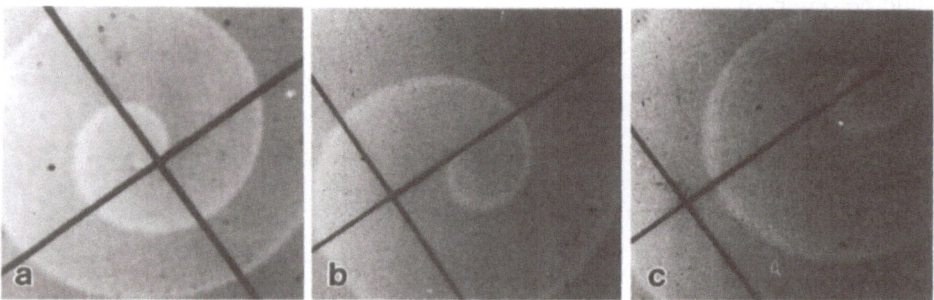

Fig. 3.28. Experimental observation of the spiral wave resonance in a chemical excitable medium with the photosensitive modification of the Belousov-Zhabotinskii reaction (from [3.61])

Figure 3.28 shows consecutive photographs taken at intervals of 10 minutes under the condition of a complete resonance (the modulation period was chosen equal to the rotation period of a spiral wave). We see that the spiral wave center slowly shifts at a constant velocity along one of the reference lines; the total shift is about 2 cm.

If the medium contains two spiral waves with opposite directions of rotation, then, by adjusting the initial phase of modulation, these two waves can be made to move towards each other and, finally, to annihilate.

Next we consider a spiral wave *drift* in inhomogeneous excitable media (see [3.60]). We assume that the critical curvature K_{cr} varies along the x-axis, but so slowly that the change of K_{cr} on the distance scale of the core radius R_0 is small:

$$|\partial K_{\text{cr}}/\partial x| \ll K_{\text{cr}}/R_0 . \qquad (3.5.16)$$

When the tip of a spiral wave moves in such an inhomogeneous medium, it passes consecutively through regions with different values of K_{cr}. Therefore, the tip experiences the periodic variation of the critical curvature:

$$K_{cr}(t) = K_{cr}\left[X_0 + R_0\cos(\omega_0 t)\right] \approx \overline{K}_{cr} + bR_0\cos(\omega_0 t) , \qquad (3.5.17)$$

where X_0 is the momentary position of the spiral wave center with respect to the x-axis,

$$\overline{K}_{cr} = K_{cr}(X_0) , \quad b = \frac{\partial K_{cr}}{\partial x}\Big|_{x=X_0} . \qquad (3.5.18)$$

Therefore, the problem of the drift is effectively the same as that of the resonance of a spiral wave in a situation when the modulation frequency coincides with the own frequency of a spiral wave. Hence, we can conclude that the spiral wave center will slowly drift along a straight line at some angle θ to the x-axis. Using (3.5.15) and (3.5.17), we find that the drift velocity is

$$V_d = \frac{(3/4)V_0}{\sqrt{1 + (D/\gamma)^2}} \cdot \frac{R_0}{K_{cr}} \cdot \left|\frac{\partial K_{cr}}{\partial x}\right| . \qquad (3.5.19)$$

The direction of the drift is determined by the angle θ_0 given by

$$\tan\theta_0 \approx -\gamma/D . \qquad (3.5.20)$$

Hence, in a medium with high inertiality (where $\gamma/D \ll 1$) spiral waves drift along the direction opposite to the gradient of K_{cr}. When inertiality is low ($\gamma/D \gg 1$), spiral waves drift in a direction which is almost orthogonal to the gradient of K_{cr}; a spiral wave drifts upwards or downwards depending on whether its rotation is clockwise or counterclockwise.

Note that the drift of a spiral wave cannot continue for an indefinitely long time. In a medium of finite dimensions, such a wave disappears when its core reaches the border. In an infinite medium, the core of a spiral wave grows as it moves into the region of smaller values of K_{cr}; the core radius becomes infinite when $K_{cr} = 0$. In this region the solution in the form of a spiral wave disappears. Note that the latter effect is not described by a simple linear theory because condition (3.5.16) is then violated.

Spiral wave motion in *anisotropic* media was investigated recently in the framework of kinematics by *Davydov* and *Zykov* [3.62].

When a spiral wave rotates not on a plane, but on a *curved surface*, it is sensitive to the local gaussian curvature of this surface (see [3.50, 63]). If the local curvature of the supporting surface is not constant but varies, spiral waves will migrate in a complicated manner.

In the effects of both drift and resonance, migrations of a spiral wave are induced by some external factors, such as a gradient in the excitability or its periodic uniform modulation. However, in a slightly more general kinematical model spiral waves can become inherently unstable with respect to migrations of their centers. This instability explains the *meandering* of spiral waves that was observed by *Winfree* (see the detailed study [3.64]).

In the simplest kinematical model we assumed that the propagation velocity of an excitation wave depends only on the local curvature of its front. This implies that the inhibitor concentration in the region which lies before the propagating wavefront does not vary. On the other hand, in Sec. 3.1 we noted that every propagating excitation wave has a tail with an increased inhibitor concentration. Therefore, as we already mentioned, the above results are valid only for sufficiently sparse wave patterns, with distances between any two consecutive waves much larger than the characteristic width of an excitation pulse, including its tail.

This limitation can be lifted, to some extent, if we introduce into this model the dependences of the propagation velocity V_0 and the critical curvature K_{cr} on the time interval T which is measured from the moment of the last passage of an excitation wave.

If the inhibitor does not diffuse, its concentration, enlarged by a propagating excitation wave, goes down to its equilibrium value independently at every point of the medium. Therefore, the residual concentration is a decreasing function of the time T that vanishes in the limit of large T.

On the other hand, the propagation velocity V_0 of excitation waves is determined by the local inhibitor concentration; this velocity is smaller for higher inhibitor concentrations. Combining these two effects, we see that V_0 should be an increasing function of T that approaches a propagation velocity of the solitary pulse in the limit $T \to \infty$.

An increased inhibitor concentration reduces the effective excitability of a medium. Hence, we can expect that it will lower the critical curvature K_{cr} as well. This implies that K_{cr} should be an increasing function of T.

The particular form of functions $V_0(T)$ and $K_{cr}(T)$ depends on a concrete reaction-diffusion model. As a possible general approximation, one can use the expressions

$$V_0(T) = V_0^{(0)}(1 - \mu/T) ,$$
$$K_{cr}(T) = K_{cr}^{(0)}(1 - \Gamma/T) .$$

$$(3.5.21)$$

Here $V_0^{(0)}$ and $K_{cr}^{(0)}$ are the corresponding quantities for a solitary propagating wave (i.e. in the limit $T \to \infty$), and μ and Γ are positive coefficients. This power-law approximation should be replaced in the limit of very long times by a law with the exponents.

Functions $V_0(T)$ and $K_{cr}(T)$ can be used instead of the parameters V_0 and K_{cr} in the equations of front kinematics. Then these equations should be supplemented by an expression for T as a function of coordinates and time.

To construct $T(x, y, t)$, we notice the time $T^*(x, y, t)$ when an excitation wave last passed through a given point with the coordinates x and y. Then the function $T(x, y, t)$ is given by a simple expression

$$T = t - T^*(x, y, t) .$$

$$(3.5.22)$$

The cartesian coordinates $x = X(l,t)$ and $y = Y(l,t)$ of a wavefront curve defined by its natural equation $K = K(l,t)$ can be found by integration of the equations

$$\frac{\partial \alpha(l,t)}{\partial l} = K(l,t) ,$$

$$\frac{\partial X(l,t)}{\partial l} = \cos\alpha , \qquad (3.5.23)$$

$$\frac{\partial Y(l,t)}{\partial l} = \sin\alpha$$

with initial conditions $\alpha(0,t) = \alpha_0(t)$, $X_0(t) = X_0(t)$, $Y(0,t) = Y_0(t)$, where α_0, X_0, and Y_0 obey (3.5.9) and (3.5.11).

For a steadily circulating spiral wave, the time interval T between any two consecutive waves is the same for all points of the medium: it is simply the rotation period of a spiral wave $T_0 = 2\pi/\omega_0$. Hence, such spiral waves are described in precisely the same manner as in Sect. 3.4. The only difference is that now we should substitute $V_0(T_0)$ and $K_{cr}(T_0)$ for V_0 and K_{cr}. To find the rotation frequency, we can use (3.4.17) which now becomes an equation for ω_0:

$$\omega_0 = \Lambda\sqrt{DV_0(T_0)}\left[K_{cr}(T_0)\right]^{3/2} . \qquad (3.5.24)$$

This equation can be easily solved, either analytically or numerically, if functions $V_0(T)$ and $K_{cr}(T)$ are known.

Note that no excitation fronts propagate inside the central core region of a spiral wave. Therefore, $T = \infty$ in the core region. On the other hand, in other parts of the medium we have $T = T_0$. Hence, a steadily rotating spiral produces a special distribution of values of T which is everywhere constant except for a circular central region of radius R_0 where T goes to infinity.

This effective nonuniformity of medium properties creates the conditions for *meandering* of spiral waves, studied in the experiments by *Jahnke* et al. [3.64], by *Müller* and *Plesser* [3.65] and by *Skinner* and *Swinney* [3.88]. Suppose we have slightly perturbed a spiral wave, so that its tip enters the core region. Then its motion is determined by the laws of kinematics with a higher propagation velocity V_0 and a higher critical curvature K_{cr} (since $T = \infty$ there). This implies that the tip will move rapidly along a trajectory of small radius. This motion will continue so long as the front does not run into its own tail, i.e. does not come into a region with a smaller value of T. When this happens, K_{cr} sharply decreases. As a result, further propagation of this portion of the front, with high curvature, is impossible. The end point shifts by a jump along the front to the point with a curvature less than the new value of K_{cr} where propagation is allowed. After that the end point will start to move at a lower velocity along the trajectory of higher radius. This will result in a slow increase in the value of T.

The above considerations indicate that a steady circulation of spiral waves can be unstable, resulting in a complicated motion of the spiral wave tip. This effect is revealed by the direct numeric integrations of the kinematics equation that were performed by *Zykov* [3.55, 67]. Figure 3.29 shows the subsequent evolution of a flat halfwave in media with different refractoriness, characterized by the parameters μ

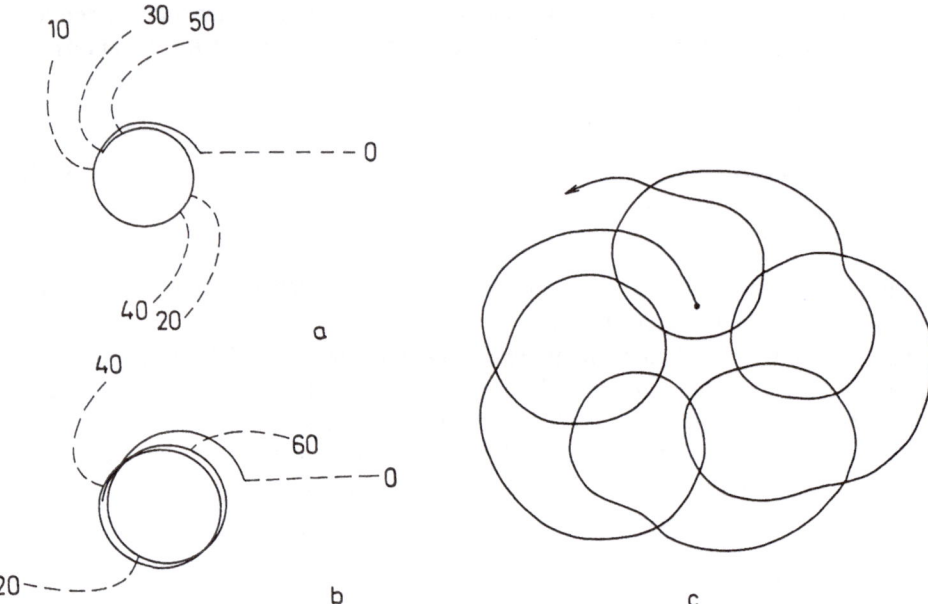

Fig. 3.29a–c. Evolution of a broken flat wave in media with different refractoriness. When refractoriness is low ($\mu = \Gamma = 1$) steady circulation is established (**a**). At a higher refractoriness ($\mu = \Gamma = 5$) small meandering of the tip around a core circle appears (**b**). At still larger refractoriness ($\mu = \Gamma = 20$), which is comparable to the rotation period $T_0 = 18$ of the spiral wave, the wave tip moves along a cycloid trajectory and the amplitude of meandering of a spiral wave is large (**c**). (From [3.67])

and Γ in (3.5.21) (in this particular simulation it was assumed that $\mu = \Gamma$). When refractoriness is low (Fig. 3.29a), a steady circulation is established. At a higher refractoriness (Fig. 3.29b), small meandering of the tip around a core circle appears. At still larger refractoriness, when μ is comparable to the rotation period T_0, the wave tip moves along a cycloid trajectory and the amplitude of meandering of the spiral wave is large (Fig. 3.29c).

To conclude our discussion of spiral waves in excitable media, it should be pointed out that the kinematic approach is not universal, even with the above generalization. In effect, it neglects the finite width of an excitation zone (where an activator is produced) and models it by a front curve. This is justified only when the rotation period T_0 of the spiral wave is much larger than the duration T_e of the excitation phase. This condition can be realized either in media of poor excitability (where K_{cr} is very small and hence $T_0 \gg T_{min}$) or in media with high refractoriness (where $T_0 \sim T_{min}$ but $T_{min} \gg T_e$).

Spiral waves are also observed in media with high excitability and moderate refractoriness. Computer simulations of full reaction-diffusion equations (see [3.19]) reveal that then the rotation period T_0 is about T_{min}. However, the phenomena in the central region are different. There is no sharply defined core in this case. The activator concentration slowly decreases towards the center. As for the inhibitor concentration, it *increases* and reaches its highest level at the central point. An excessive inhibitor concentration raises the excitation threshold in the core region.

In this way an effective obstacle to propagating excitation waves is produced. In this case the inverse gradient of the inhibitor stabilizes the rotation of spiral waves and no meandering is observed.

3.6 Wave Patterns in Three Dimensions

If we take a rotating spiral wave on a plane and continue it straight upwards, a three-dimensional wave pattern which is called a *scroll wave vortex* is obtained (Fig. 3.30). The properties of scrolls are analogous to those of spirals. While the spiral wave rotates around some point on a plane, the rotation centers of a scroll wave fill an entire line in the space, which is called a vortex *filament*.

The vortex filament need not necessarily be a straight line. For instance, it can be bent and linked into a circle (Fig. 3.31). By means of such a transformation, a *scroll ring* is created (Fig. 3.32).

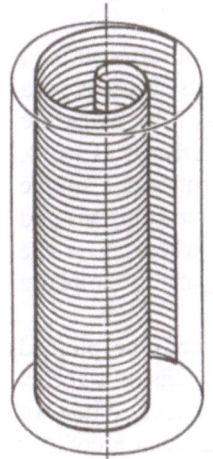

Fig. 3.30. Straight scroll vortex

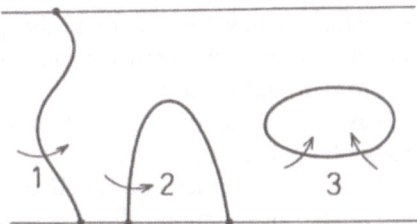

Fig. 3.31. Possible deformations of the vortex filament. Arrows indicate the direction of rotation

Scroll rings are local wave sources. Far from the center, they have the same pattern of expanding spherical waves as a three-dimensional pacemaker.

Scrolls and scroll rings are only the simplest examples of self-supporting wave patterns in three dimensions. Much more complicated patterns can be conceived. If a straight scroll wave is twisted along its vertical axis, the resulting pattern is a *twisted scroll* (Fig. 3.33). When the excitation waves from such vortex cross the surface of an imaginary cylinder surrounding the vortex filament they leave on it a screw line. By bending the filament of a twisted scroll wave and linking it into a circle, we can create an object that can be called a twisted scroll ring.

In a similar way one can go on to construct much more complicated pattterns, such as linked and/or knotted vortex rings, etc. However, there are some topological limitations, found by *Winfree* and *Strogatz* [3.68, 69], that significantly reduce the

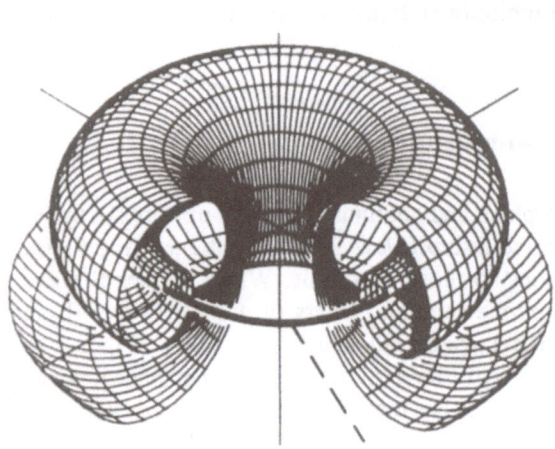

Fig. 3.32. Scroll ring

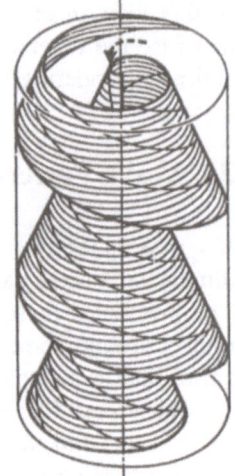

Fig. 3.33. Straight twisted scroll

number of permitted three-dimensional wave patterns. For example, it turns out that we cannot link into a chain any two simple (untwisted) scroll rings. Moreover, a solitary twisted scroll ring is also impossible (if we try to construct it by using the above mentioned procedure, another vortex will be simultaneously created as well).

Let us indicate how such topological restrictions arise. First, we consider a simple vortex and surround its filament by a closed contour. Note that, after going around the filament along this contour, the phase is increased by $2\pi N$ where N is the number of arms in the vortex. If the contour did not surround the filament, the phase increment would be zero.

Now it is easy to see that a twisted scroll ring should necessarily be pierced through by a filament of another vortex. Let us encircle a symmetry axis of a vortex that is twisted M times (M is called a *total twisting*) by a closed contour. When we walk around this contour, the phase is incremented by $2\pi M$. Thus, such a contour should be pierced by a filament of an M-armed vortex or by several vortex filaments with a total topological charge M[6].

Multi-armed twisted vortexes can be produced by twisting certain multi-armed straight vortexes along their axes. If we bend the filament of a multi-armed vortex into a circle, we create a multi-armed twisted scroll ring. Note that the total twisting of a multi-armed scroll ring can acquire fractional values. Since it is permitted to glue one arm of a vortex to another, the total twisting can take values $M = m/N$, where m is a certain integer and N is the number of arms in this vortex.

The phase change after traversing a contour that surrounds the vertical axis of the ring is equal to NM. By repeating the previous arguments, we can show that

---

[6] The concept of a topological charge for simple vortexes was proposed by *Zeldovich* and *Malomed* [3.70]. The topological charge of a spiral wave in a two-dimensional medium is equal to the change of phase (divided by 2π) after passing in the clockwise direction along any closed contour which surrounds the core of this spiral wave. For spiral waves that rotate in the opposite direction, the sign of the topological charge is reversed.

such a ring should be pierced through by a vortex with NM arms (or by several vortexes with the total topological charge NM).

The filament of a vortex that pierces a twisted scroll ring can also be bent into a circle. Thereby a pair of twisted scroll rings, linked with each other, is created.

The number of arms and the total twisting of two linked vortexes are not arbitrary; they should satisfy a certain relationship. Suppose that N_1 is the number of arms and M_1 is the total twisting of the first ring. Then, as follows from the above considerations, the number of arms in the second ring should be equal to $N_2 = N_1 M_1$. In a similar way we find that the number of arms in the first ring is $N_1 = N_2 M_2$ where M_2 is the total twisting of the second ring. These two equations are consistent only if

$$M_1 M_2 = 1 , \quad N_1/N_2 = \sqrt{M_1/M_2} . \tag{3.6.1}$$

It follows from (3.6.1) that, if the first of the two rings is twisted M times, then the second one is twisted around only by $1/M$ of a full rotation. Besides, the second ring has M times more arms than the first one.

Relations (3.6.1) impose severe restrictions on possible ring links. They imply, for instance, that we cannot link two simple (untwisted) scroll rings, or two one-armed scroll rings each twice twisted.

Vortex filaments can be tied into *knots* as well. The resulting wave patterns turn out to be extremely complicated. Their classification and selection of the permitted knots require advanced methods of mathematical topology (see [3.69]).

Vortexes in three-dimensional active media were observed in the Belousov-Zhabotinskii reaction by *Winfree* [3.36] and by *Welsh* et al. [3.71]; they were also studied in numerical simulations [3.72-75].

Computer simulations show that scroll rings are usually unstable. As a rule, they slowly shrink at a rate inversely proportional to the ring radius and finally disappear. In some cases, however, inflation of a scroll ring is observed: its radius increases and it transforms into a straight rotating scroll.

Similarly to spiral waves, scrolls can be sparse or tight, depending on the values of the parameters which characterize the effective excitability of the medium. When a scroll ring is tightly wound, it emits waves at a period close to T_{min}, so that the front of the following wave almost runs into the tail of the preceding one.

Below we briefly outline the theoretical description of sparse three-dimensional wave patterns based on the kinematics of front surfaces that was proposed by *Brazhnik* et al. [3.45, 55, 76].

The kinematical approach developed for two-dimensional excitable media can easily be generalized to three dimensions. We assume now that a wave is completely defined by its oriented front surface. Any surface in a three-dimensional space is characterized, at any given point, by two principal curvature radii R_1 and R_2. For small curvatures, the normal propagation velocity of a small surface segment depends linearly on the sum of the two principal curvatures, i.e. on the doubled *mean* curvature $2H = 1/R_1 + 1/R_2$, so that

$$V = V_0 - 2DH , \tag{3.6.2}$$

where D is a certain proportionality factor.

A front surface can sprout or contract at its free edge (where it is cut). The velocity G of such tangent motion depends not only on the mean front curvature H at the edge but also on the geodetic curvature κ of the edge line itself. For small curvatures, we have in the linear approximation

$$G = -\gamma_1(2H - K_{cr}) - \gamma_2\kappa . \tag{3.6.3}$$

Here γ_1 and γ_2 are positive coefficients; the geodetic curvature of the edge line is taken to be positive if the edge line is convex with respect to the front surface.

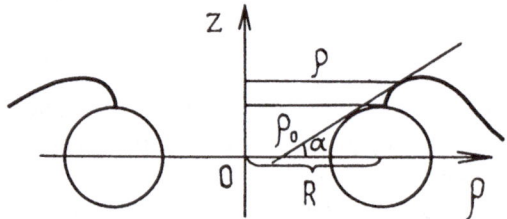

Fig. 3.34. The meridian section of a scroll ring

We can use the framework of a kinematic theory to consider the evolution of a simple (untwisted) scroll ring. Let us introduce a cylindrical coordinate system (z, ϱ, φ) with a z-axis that coincides with the symmetry axis of the ring. Because of the rotation symmetry, it suffices to analyze the evolution of a curved line (the scroll ring *meridian*) that is produced by the cross section of the ring in a plane $\varphi = \mathrm{const}$ (Fig. 3.34). Since this meridian curve is flat, the kinematic equation for its evolution coincides with (3.4.9). However, velocities V and G are given now by (3.6.2) and (3.6.3). The mean curvature H of a scroll ring is

$$H = (1/2)(K - \varrho^{-1}\sin\alpha) , \tag{3.6.4}$$

and the geodetic curvature of the edge line is

$$\kappa = -\varrho_0^{-1}\cos\alpha_0 . \tag{3.6.5}$$

Here K is the local curvature of the meridian, ϱ is the distance from the z-axis, α is the angle between the tangent to the meridian and the ϱ-axis (see Fig. 3.34), α_0 is this angle at the end point of the meridian, and ϱ_0 is the distance from the z-axis to the end point.

When the quasistationary approximation is applicable, the form of a front surface near its free edge quickly adjusts to the mean curvature of this surface near the edge and the (geodetic) curvature of the edge line at a given time moment. Therefore, it is sufficient to consider only the motion of the end point of the meridian curve (i. e. of the point where the scroll edge crosses the meridian plane). This motion is described (see [3.45]) by equations

$$\dot\varrho_0 = -V(0)\sin\alpha_0 - G\cos\alpha_0 ,$$
$$\dot z_0 = V(0)\cos\alpha_0 - G\sin\alpha_0 , \tag{3.6.6}$$

$$\dot{\alpha}_0 = -D(\partial K/\partial l)_0 - (DK_0/\varrho_0)\cos\alpha_0 - (D/2\varrho_0^2)\sin 2\alpha_0 + GK_0 , \qquad (3.6.7)$$

$$\dot{K}_0 = -\left[G - (D/\varrho_0)\cos\alpha_0\right] - (DK_0/2\varrho_0)(3\cos 2\alpha_0 + 1) , \qquad (3.6.8)$$

where

$$(\partial K/\partial l)_0 = \Lambda\sqrt{V_0/D}K_0^{3/2} . \qquad (3.6.9)$$

Together with (3.6.2–5), these equations determine the motion of the meridian end point, and hence time evolution of the scroll ring.

We assume that the scroll ring radius R is much larger than the thickness of its filament core, which is about the diameter of a spiral wave core. It can easily be verified that for $R \to \infty$, i.e. for a straight cylindrical scroll, (3.6.6–8) have a solution for which the rotation frequency is the same as that of a spiral wave (i.e. $\alpha_0 = \omega_0 t$), and K_0 is equal to K_{cr}. Within the linear in $1/R$ approximation we can replace ϱ_0 by R and put $\alpha_0 \approx \omega_0 t$ in the right sides of (3.6.7) and (3.6.8). After this transformation we are left with equations that are very similar to those which describe the resonance of spiral waves (in a special case when the modulation frequency coincides with the rotation frequency of a spiral wave). Therefore, we can expect that the center of the vortex filament will slowly drift in the meridian plane at a velocity proportional to the effective modulation amplitude, i.e. to the inverse radius $1/R$.

The detailed calculation [3.45] shows that, in the linear in $1/R$ approximation, the scroll ring radius changes in time according to the equation

$$\dot{R} = -\frac{D}{R}\left(1 - \frac{3}{4\beta^2}\cdot\frac{q_1^2 + q_2}{q_1^2 + 1}\right) , \qquad (3.6.10)$$

where $\beta = (DK_{cr}/V_0)^{1/2} \ll 1$ and where we have introduced the notations $q_1 = \gamma_1/D$ and $q_2 = (\gamma_2 - D)/D$. At the same time, the ring drifts in the vertical direction (i.e. along its symmetry axis) with the velocity

$$\dot{Z}_0 = -\frac{D}{R}\left(\frac{q_2}{2\Lambda\beta} + \frac{3}{4\beta^2}\cdot\frac{q_1 q_2}{q_1^2 + 1}\right) . \qquad (3.6.11)$$

If a scroll rotates in the opposite direction, the direction of its drift along the symmetry axis is also reversed.

Therefore, a scroll ring is almost always unstable. It either shrinks (if $dR/dt < 0$) and disappears, or inflates (if $dR/dt > 0$). Collapse or inflation are accompanied by a drift of the ring along its symmetry axis.

There is, however, a narrow interval of the parameters q_1 and q_2 where dR/dt is very small and changes its sign (see (3.6.10)). Within this interval we should take into account in the expression for dR/dt the terms of higher orders in $1/R$. A careful examination reveals[7] that the terms of order $1/R^2$ vanish after averaging

---

[7] This analysis was performed by *Khrustova* [3.77].

over the rotation period, whereas the cubic term $1/R^3$ enters into the expression for dR/dt with a positive coefficient. Therefore, if in the linear approximation a scroll ring is shrinking but the speed of shrinking is very small, this nonlinear positive term will prevent collapse and stabilize the scroll ring.

When $q_1^2 \gg q_2$, the radius of a stable scroll ring is given by

$$R = R_0 \frac{\sqrt{(13/32)\Lambda\beta q_1}}{\sqrt{(q_1/\beta)^2 - 3/4}} \qquad (3.6.12)$$

where $R_0 = V_0/\omega_0$ is the filament radius of the scroll ring which coincides, in this approximation, with the spiral wave core radius in the same excitable medium.

Equation (3.6.12) implies that stable scroll rings exist only for $q_1^2 > (3/4)\beta^2$. The radius of a stable scroll ring becomes infinite at $q_1^2 = (3/4)\beta^2$; it rapidly diminishes for larger values of q_1. Equation (3.6.12) remains valid while $R \gg R_0$. It can be expected that the solution for a stable scroll ring is lost when R is about R_0.

Note that, according to (3.6.11), stable scroll rings slowly drift with velocity about $\beta(D/R)$ along their symmetry axes.

Stable scroll rings were observed in a numeric simulation [3.45, 46] for a particular model of an excitable medium.

It was already mentioned that, far from its center, a scroll ring is indistinguishable from a spherical pacemaker. There is, however, a principal difference. A genuine pacemaker exists only in inhomogeneous media, i.e. if the properties of a certain local group of elements are modified so that they begin to generate periodic oscillations. A scroll ring can exist in a homogeneous medium. Moreover, it does not rest: its center moves along the internal symmetry axis.

Although the above results were found for circular scroll rings, they can also be used to describe in a simplest approximation the time evolution of any vortex with a slightly deformed filament. Indeed, if we consider a small section of such a vortex, it will look like a part of a scroll ring of the same radius.

a   b

Fig. 3.35a,b. Evolution of a perturbation in the filament form of a straight scroll vortex in the medium with a higher (a) and a lower (b) excitability

Suppose that initially we had a straight scroll and then we have locally perturbed its filament (Fig. 3.35a). If the parameters of a given medium are such that scroll rings shrink there, this perturbation will diminish in time. Hence, we can say that a vortex filament possesses in this case a certain "elasticity", tending to shorten its length. In the opposite case, when a scroll ring inflates, any protrusion of a filament will increase in time, leading to filament elongation (Fig. 3.35b). Consequently, a straight cylindrical scroll turns out to be unstable with respect to small local deformations of its filament. This instability should lead to the formation of very complex (and, possibly, even chaotically organized) wave patterns in such media. The filament dynamics of (both simple and twisted) scrolls was investigated by *Keener* [3.78].

A resonance of scroll rings was studied by *Abramychev* et al. [3.79]. By applying periodic modulation of the medium excitability and adjusting the amplitude and the phase of such modulation, it is possible to stabilize inflating scroll rings and control the velocity of their drift along a central axis.

To conclude this section, we discuss properties of twisted vortexes. For a cylindrical twisted vortex (Fig. 3.33), the front edge curve represents a screw line that is wound onto a cylinder (the vortex core filament). The local twistedness of such a vortex is characterized by a parameter $W = 2\pi/h$, where h is the step of a screw line. This parameter can be positive or negative, depending on whether the screw is right or left.

The kinematic theory predicts that the rotation frequency of a twisted scroll should be higher than that of a simple one. For small values of W, the rotation frequency of a twisted scroll wave is found by *Brazhnik* et al. [3.80, 81] to be

$$\omega = \omega_0 \left[1 + (3/2)(\gamma_2/\gamma_1)(|W|/K_{cr})\right] . \tag{3.6.13}$$

Computer simulations (see [3.82, 83]) indicate that, for every excitable medium, there is a maximal permitted value of W; if we try to twist a vortex further it will be torn apart. A maximally twisted vortex sends waves at a period about T_{min}, i.e. at a period close to the minimal possible interval between two consecutive excitation waves.

If we twist a vortex nonuniformly, so that $W = F(z)$, the "waves of twistedness" will start to propagate. In the first approximation, the pulses of twistedness propagate without changing their form at a speed (see [3.80, 81])

$$u_W = (3/2)(\gamma_2/\gamma_1)(\omega_0/K_{cr}) . \tag{3.6.14}$$

The direction of propagation is determined by the sign of W. When two propagating pulses collide, an analog of a shock structure is produced (this discontinuity is smoothed if higher spatial derivatives of W are included in the evolution equation, see [3.84]). Note also that, at higher values of W, a screw deformation of the scroll filament can develop because of the interaction between the twistedness and the filament form (see [3.78]).

When a cylindrical twisted scroll is created in a finite volume of an excitable medium and its filament comes out onto the boundary surface, the pattern is unstable. It twists off and transforms into a simple straight scroll. However, stable twisted scrolls are naturally formed in *nonhomogeneous* excitable media, as it was shown by *Panfilov* et al. [3.82].

Suppose that an excitable medium consists of two parts. In the upper part, spiral waves (and untwisted scrolls) have rotation period T_1, whereas the rotation period in the lower part is T_2, and $T_2 > T_1$. Then, if we create initially a straight simple scroll with the filament orthogonal to the boundary between these two parts, it will become twisted in its lower part (see [3.47]). The degree of twistedness (i.e. the parameter W) can be found from a simple condition: the rotation period T of the twisted part of the vortex should be the same as the rotation period T_1 of a simple untwisted vortex in the upper part of the medium (remember that twisting increases the rate of rotation).

This example shows that scroll waves in nonhomogeneous media with slowly varying parameters should, as a rule, be twisted. Twisted linked scroll rings were studied in a computer simulation by *Nandapurkar* and *Winfree* [3.85].

4. Oscillatory Media

Oscillatory media represent a continuous limit of a large population of self-oscillating elements, with weak interactions between the neighbors. Since interactions between the neighboring elements are weak, they cannot significantly change the amplitude and the form of individual oscillations. Therefore such interactions are principally manifested in changes of the oscillation phases. This notion allows one to construct an approximate description of processes in oscillatory media in terms of phase variables.

Depending on the properties of a medium, oscillations can either synchronize or desynchronize with time; in the latter case a turbulent regime is established.

By creating a constant phase gradient along some direction, one can produce a plane wave with a certain wavelength. If a medium includes an inhomogeneity, i. e. a small region where the local oscillation frequency is higher than in the rest of the medium, this region gives birth to a pacemaker which represents a source of periodic concentric waves. When two such sources operate in the same medium, the higher-frequency pacemaker suppresses the pacemaker with the lower frequency.

In contrast to pacemakers, spiral waves in oscillatory media exist even in absence of any inhomogeneities. The rotation frequency of such waves is again, as it was in excitable media, a unique property of a given medium. Because of the phase singularity in the center of a spiral wave, these patterns cannot be consistently described within the phase dynamics approximation; the complete dynamical equations of a medium should be used to determine, for instance, the rotation frequency of such waves.

4.1 Phase Dynamics

Let us consider an active element that maintains periodic (but possibly anharmonic) oscillations. Although the amplitude, the frequency, and the form of established self-oscillations are uniquely determined for a given element, their phase Φ remains arbitrary. If a small perturbation is introduced, the initial form and the amplitude of oscillations will be recovered after some relaxation time τ_{rel}. However, the phase Φ will not return to its original value: as a consequence of a perturbation it will get a small correction.

Suppose we have a distributed active medium where every small element is self-oscillatory. As an approximation, it can be thought of as a regular network of oscillating elements. We assume that interactions in this oscillatory medium are local and that they have a diffusional character. Interaction is absent when the momentary states of neighboring elements are the same; generally it depends on the difference of state variables in the neighboring elements.

Obviously, this *oscillatory medium* can maintain uniform synchronous self-oscillations with a constant phase. In such a regime, the amplitude and the form of oscillations are the same as those of an isolated individual element.

If the phase Φ varies smoothly across the medium (with a large characteristic length L), the evolution of such phase distribution is slow. The larger the length scale L, the smaller is the phase gradient and the weaker should be the interaction of any two neighboring elements.

On the other hand, amplitude perturbations are quickly damped, independent of their characteristic length scales. This damping is determined by the properties of an individual element and occurs in a time of about τ_{rel}.

Therefore, for distributions with large length scales, characteristic evolution times of phases and amplitudes are sharply different. Since the amplitude relaxation time is relatively short, amplitudes adjust adiabatically to a local phase gradient $\nabla\Phi$. When this gradient is small, the deviations of the amplitudes from their equilibrium value are small too.

The dynamics of smooth phase distributions should be governed by an effective equation that does not involve the amplitudes of oscillations. This equation can be found by rather general arguments.

In an isotropic active medium, the rate of change $\partial\Phi/\partial t$ of the phase cannot depend on the direction of the phase gradient $\nabla\Phi$. Therefore, for sufficiently small gradients, decomposition of $\partial\Phi/\partial t$ in terms of $\nabla\Phi$ should begin with a quadratic term, i.e.

$$\frac{\partial\Phi}{\partial t} = a(\nabla\Phi)^2 + \dots . \tag{4.1.1}$$

According to (4.1.1), phase variations with a length scale L have evolution rates of the order $1/L^2$. Note that, along with the gradient term, the right hand side of (4.1.1) can also include some terms with higher order spatial derivatives. We should keep only those such terms which lead to contributions of order $1/L^2$. It can easily be verified that there is only one such term, proportional to the phase Laplacian $\Delta\Phi$.

Thus, a dynamical evolution equation for smooth phase distributions should have the form

$$\frac{\partial\Phi}{\partial t} = a(\nabla\Phi)^2 + b\Delta\Phi , \tag{4.1.2}$$

where a and b are coefficients that have the dimensions of a diffusion constant (i.e. $[a] = [b] = cm^2/s$). The specific values of these coefficients depend on a particular model of an oscillating active medium: they can be found from the full dynamical equations (see the examples below).

Note that, like the kinematical equations for excitable media, the phase dynamics equation (4.1.2) is *universal* in the sense that every oscillatory medium is described here by only two parameters a and b, irrespective of the details of interactions between the elements or of a particular form of oscillations.

It follows from (4.1.2) that a characteristic time scale for the evolution of phase distributions with a length scale L is about $\tau_L \sim L^2/b$. Since the derivation of (4.1.2) was performed under the assumption that characteristic times for phases are large compared to the amplitude relaxation time τ_{rel} for an individual oscillator, this equation is valid only for very smooth phase distributions with length scales L satisfying a condition

$$L \gg \sqrt{b\tau_{rel}} .\qquad(4.1.3)$$

This condition limits the applicability of the phase dynamics equation.

The equation of phase dynamics can be rigorously derived from the equations of a reaction-diffusion system, provided we know the solution for homogeneous self-oscillations in such a system. This derivation, which represents a variant of the *Whitham method* (see [4.1]) in the theory of nonlinear waves in conservative media, can be found in the book by *Kuramoto* [4.2]. The related questions were discussed by *Howard* and *Kopell* [4.3, 4], *Kuramoto* [4.5] and *Sivashinsky* [4.6]; for a recent review see [4.7].

Following *Kuramoto* and *Tsuzuki* [4.8], we now derive the phase dynamics equation in the important special case of *quasiharmonical* oscillatory media.

Suppose that an oscillatory medium is described by two variables $u = u(r, t)$ and $v = v(r, t)$ which can be combined into a single complex-valued variable $\eta(r, t) = u(r, t) + iv(r, t)$. We assume that evolution of η obeys the equation

$$\frac{\partial \eta}{\partial t} = \lambda(\varrho)\eta - i\omega(\varrho)\eta + (D_1 + iD_2)\Delta\eta ,\qquad(4.1.4)$$

where λ and ω are functions of the amplitude $\varrho = |\eta|$; function $\lambda(\varrho)$ is positive at $\varrho < \varrho_0$, vanishes at $\varrho = \varrho_0$, and becomes negative at $\varrho > \varrho_0$.

A mathematical model of an oscillatory medium given by (4.1.4) is called a $\lambda - \omega$ model. If the complex coefficient $D = D_1 + iD_2$ that plays the role of a diffusion constant in (4.1.4) is equal to zero, such a medium breaks down into a set of disjoint oscillatory elements. In this case any individual element of the medium performs harmonic oscillations

$$\eta(t) = \varrho_0 \exp\left[-i(\omega_0 t + \Phi)\right]\qquad(4.1.5)$$

with an amplitude ϱ_0 determined by the condition $\lambda(\varrho_0) = 0$ and a frequency $\omega_0 = \omega(\varrho_0)$. The initial phase Φ of these oscillations remains arbitrary.

Small amplitude deviations $\delta\varrho = \varrho - \varrho_0$ for an individual element of the medium relax according to equation (cf. (4.1.4))

$$\delta\dot{\varrho} = \varrho_0 \lambda'(\varrho_0) \delta\varrho .\qquad(4.1.6)$$

Therefore, the amplitude relaxation time is

$$\tau_{rel} = |\varrho_0\lambda'(\varrho_0)|^{-1}\qquad(4.1.7)$$

Let us consider now a general case with $D \neq 0$. If we define a local amplitude $\varrho(r, t)$ and a local phase $\Phi(r, t)$ by

$$\eta(\mathbf{r}, t) = \varrho(\mathbf{r}, t) \exp\left[-i\left(\omega_0 t + \varPhi(\mathbf{r}, t)\right)\right] , \tag{4.1.8}$$

they satisfy equations

$$\frac{\partial \varrho}{\partial t} = \lambda(\varrho)\varrho + D_1 \Delta\varrho - D_1 \varrho(\nabla\varPhi)^2 + D_2 \varrho \Delta\varPhi + 2D_2 \nabla\varrho \nabla\varPhi , \tag{4.1.9}$$

$$\frac{\partial \varPhi}{\partial t} = \left[\omega(\varrho) - \omega_0\right] + 2(D_1/\varrho)\nabla\varrho \nabla\varPhi + D_1 \Delta\varPhi$$
$$- (D_2/\varrho)\Delta\varrho + D_2 (\nabla\varPhi)^2 . \tag{4.1.10}$$

Only smooth distributions with large length scales L are considered here. Hence, it can be expected that all oscillation amplitudes are close to ϱ_0, i.e. $\delta\varrho/\varrho_0 \ll 1$. Moreover, amplitude deviations $\delta\varrho(\mathbf{r}, t)$ adjust adiabatically to momentary values of $\nabla\varPhi$ and $\Delta\varPhi$. With the accuracy of the terms of order $1/L^2$ we have

$$\delta\varrho = \varrho_0 \tau_{\mathrm{rel}}\left[D_2 \Delta\varPhi - D_1(\nabla\varPhi)^2\right] . \tag{4.1.11}$$

Indeed, by substituting (4.1.11) into (4.1.9), it can be verified that we have neglected here only the terms of higher orders in $1/L$.

To proceed with the derivation, we substitute $\varrho = \varrho_0 + \delta\varrho$ into equation (4.1.10) for $\varPhi$ and keep only the terms of order $1/L^2$, thus obtaining

$$\frac{\partial \varPhi}{\partial t} = D_1 \Delta\varPhi + D_2 (\nabla\varPhi)^2 + \omega'(\varrho_0) \delta\varrho , \tag{4.1.12}$$

where $\delta\varrho$ is given by (4.1.11). In explicit form this equation reads

$$\frac{\partial \varPhi}{\partial t} = \left(-D_1 \frac{\omega'(\varrho_0)}{|\lambda'(\varrho_0)|} + D_2\right)(\nabla\varPhi)^2 + \left(D_1 + \frac{\omega'(\varrho_0)}{|\lambda'(\varrho_0)|} D_2\right)\Delta\varPhi , \tag{4.1.13}$$

which coincides with (4.1.2).

A special case of a $\lambda - \omega$ model is a *generalized Ginzburg-Landau equation*[1]

$$\frac{\partial \eta}{\partial t} = (\alpha_1 - i\alpha_2)\eta - (\beta_1 + i\beta_2)|\eta|^2\eta + (D_1 + iD_2)\Delta\eta . \tag{4.1.14}$$

In this case we have $\lambda(\varrho) = \alpha_1 - \beta_1 \varrho^2$ and $\omega(\varrho) = \alpha_2 + \beta_2 \varrho^2$. Therefore, according to (4.1.13), the coefficients of the phase dynamics equation are

$$a = -(\beta_2/\beta_1)D_1 + D_2, \; b = (\beta_2/\beta_1)D_2 + D_1 . \tag{4.1.15}$$

Let us turn now to the analysis of processes described by the phase dynamics equation (4.1.2). First, we note that by the transformation

$$\varPhi(\mathbf{r}, t) = (b/a)\ln Q(\mathbf{r}, t) \tag{4.1.16}$$

(4.1.2) can be reduced to a linear equation

---

[1] The usual time-dependent Ginzburg-Landau equation (see [4.9]) describes the dissipative dynamics of order parameters near second-order phase transitions, such as a superconductive transition in metals; it has $\alpha_2 = \beta_2 = D_2 = 0$. The analogy between various wave patterns in oscillatory media and vortexes in superconductors is discussed in [4.10, 11].

$$\frac{\partial Q}{\partial t} = b \, \Delta Q \ . \tag{4.1.17}$$

Therefore, a decisive role in the evolution of such a system is played by the sign of the coefficient b. If this coefficient is positive, (4.1.17) is just a standard diffusion equation. Therefore, every *localized* perturbation in Q (and, respectively, in Φ) with a characteristic length scale L spreads out and disappears in a time of about L^2/b. This results in recovery of the initial uniform distribution, for which Q and Φ did not depend on the coordinates. Therefore, oscillations tend to become synchronized if $b > 0$.

An opposite situation is found if b is negative. In this case an initial localized perturbation does not spread, but collapses and simultaneously increases in magnitude with a characteristic time L^2/b. Consequently, a uniform regime with synchronous oscillations turns out to be unstable with respect to small perturbations. Fluctuations grow, and the most rapid growth is typical for the shortest-scale fluctuations. This results in the establishment of a *chaotic* or turbulent regime (see e. g. [4.2, 12]). Note that, since the evolution of local phase perturbations is accompanied by a decrease of characteristic length scales, after some time has passed condition (4.1.3) becomes violated and (4.1.2) is no longer applicable. Within the phase description, a turbulent regime is manifested in creation of multiple phase singularities. A complete analysis of this regime can only be carried out by using the full dynamical equations of a distributed oscillatory medium. Therefore, the properties of the turbulent regime are not universal; they are very sensitive to the specific form of equations that describe an oscillatory active medium.

4.2 Plane Waves

A plane wave is observed in an oscillatory medium if there is a constant phase gradient, i. e. if

$$\nabla \Phi = -\boldsymbol{k} = \text{const} \ . \tag{4.2.1}$$

Because of such a phase gradient, the oscillations of neighboring elements along the direction of $\boldsymbol{k}$ are performed with a constant phase shift, which gives the appearance of wave propagation. "Waves" running along decorative garlands of electric bulbs, switched on in turn one after another for a short time, are an effect of similar origin.

Hence, to establish a plane wave in a medium one has to create a certain initial phase gradient. The smaller this gradient, the larger is the propagation velocity of the wave.

Interactions between medium elements (due to diffusion) do not play any significant role for fast waves with very small phase gradients[2]. However, for slower waves with higher phase gradients k, interaction effects become essential. Generally, if a plane wave with a wavenumber k is propagating in the medium, its elements

---

[2] In the experiment of *Kopell* and *Howard* [4.13] an oscillating chemical solution with the BZ reaction, where fast plane waves were induced, was separated into two parts by a thin plexiglass plate. This had no effect on the "propagation" of waves, which "passed" through the plate without any distortion.

perform oscillations at a frequency that is different from the frequency ω_0 of uniform self-oscillations.

Propagation of plane waves can be analyzed in the framework of the phase dynamics equation (4.1.2). Such a wave corresponds to a special solution of this equation:

$$\Phi = -\boldsymbol{k}\boldsymbol{r} + ak^2 t \ . \tag{4.2.2}$$

The last term describes a nonlinear shift of the oscillation frequency in a propagating wave:

$$\omega = \omega_0 + ak^2 \ . \tag{4.2.3}$$

The sign of a determines whether the oscillation frequency increases or diminishes for higher wavenumbers k. The propagation velocity of such a wave is

$$c = \omega_0 / k + ak \ . \tag{4.2.4}$$

Within an approximation of phase dynamics given by (4.1.2), all plane waves are stable when $b > 0$. If $b < 0$, a special solution in the form of a plane wave still exists, but is not now of any particular interest since it is unstable with respect to growth of small local perturbations leading to establishment of a chaotic space-time regime.

Waves with a small spatial period, for which $k \gtrsim (b\tau_{\mathrm{rel}})^{-1/2}$, are not described by the phase dynamics equation (4.1.2). Their analysis should be carried out in the framework of the full equations of an oscillatory medium. Such waves usually lose their stability at $k \sim (b\tau_{\mathrm{rel}})^{-1/2}$. Properties of plane wave solutions in various reaction-diffusion systems have been investigated, for instance, in [4.14–17].

Let us now discuss what happens when two plane waves with different wave numbers collide. For simplicity, we assume that these waves move towards one another along the x-axis, so that $\Phi(x) \to -k_1 x$ as $x \to -\infty$ and $\Phi(x) \to k_2 x$ as $x \to +\infty$.

There is a special solution of (4.1.2) that describes two colliding waves, i.e.

$$\begin{aligned} Q(x,t) = &A_1 \exp\{-(a/b)k_1 x + (a^2/b)k_1^2 t\} \\ &+ A_2 \exp\{(a/b)k_2 x + (a^2/b)k_2^2 t\} \ , \end{aligned} \tag{4.2.5}$$

where coefficients A_1 and A_2 are determined by the initial conditions. This can be seen if we recall that

$$\Phi(x,t) = (b/a) \ln Q(x,t) \ . \tag{4.2.6}$$

Since Q is given by a sum of two different exponential terms, almost everywhere on the x-axis only one of these two terms plays a dominant role. Outside of a narrow region of width about

$$\delta x \sim \frac{b}{a(k_1 + k_2)} \tag{4.2.7}$$

with the center at the point

$$X(t) = a(k_1 - k_2)t , \tag{4.2.8}$$

the second term dominates for $x > X(t)$ and the first term dominates for $x < X(t)$. Since the first term in (4.2.6) corresponds to a wave that propagates in the positive direction of the x-axis and the second term describes a wave that moves in the opposite direction, we can conclude that these two colliding waves quench one another (i. e. annihilate) within a narrow layer of width δx, which is slowly moving (see (4.2.8)) at a constant velocity $V_a = a(k_1 - k_2)$ along the x-axis.

Note that, by taking into account the dispersion relation $\omega = \omega_0 + ak^2$ of periodic plane waves, the expression for the velocity of motion of the interface between regions occupied by these two waves can also be written in the form

$$V_1 = \frac{\omega_1 - \omega_2}{k_1 + k_2} . \tag{4.2.9}$$

Here ω_1 and ω_2 are the oscillation frequencies of the first and the second waves, respectively.

Therefore, as time elapses, this interface moves farther into the region occupied by the wave with the lower frequency. In other words, the region with high-frequency oscillations gradually grows and finally takes over the region where the oscillations had a lower frequency.

4.3 Pacemakers

Above we considered only the phenomena in homogeneous oscillatory media where properties of all elements are identical. Now we can consider the effects of local inhomogeneities. If the oscillation frequency within some small region is considerably larger than in the rest of the medium, such a region becomes a source of expanding concentric wave, or a *pacemaker*.

Pacemakers in the Belousov-Zhabotinskii chemical reaction (which are called "target patterns") were discovered by *Zaikin* and *Zhabotinskii* [4.18]. Further experiments (see, e. g., [4.19]) proved that stable target patterns are produced by impurities, which change local properties of the BZ solution and either increase the local oscillation frequency or transfer the solution from the excitable to the oscillatory regime. The observed properties of target patterns in the BZ reaction are well explained by the theory of *Tyson* and *Fife* [4.20, 21] which is based on this assumption.

The problem of the possible existence of autonomous pacemakers in homogeneous oscillatory media (i. e. without any impurities) has been actively discussed in the last two decades. Several approximate solutions [4.22–24] that describe such wave patterns were proposed. However, the stability of the proposed solutions remains uncertain. Numeric simulations usually show that the pacemakers in homogeneous media are unstable: they disappear after a finite number of emitted waves. The intrinsic instability of such patterns seems likely because, in contrast to spiral waves, there are no topological conservation laws that could prevent their continuous transformation into uniform oscillations.

The "topological charge" of a planar pattern (see [4.11, 4.25]) is equal to the phase increment, divided by 2π, after traversing a closed contour which surrounds the center of a pattern. Thus, any single-armed spiral wave has a unit topological charge (its sign depends on the rotation direction). Since the topological charge of the uniform state is obviously zero, single spirals cannot spontaneously emerge or disappear. On the other hand, the topological charge of a pacemaker emitting circular waves is also zero, and hence there are no topological limitations for its slow disappearance[3].

Below we consider, following *Kuramoto* [4.2], only long wavelength pacemakers which can be described within the phase dynamics approximation. In this case it can be shown that no autonomous pacemakers are possible, i.e. all the pacemakers are associated with some inhomogeneities.

Suppose that the oscillation frequency of individual elements is given by a function $\omega(r)$ that approaches ω_0 as $r \rightarrow \infty$ and exceeds ω_0 inside a region with a characteristic size r_0. Within the phase dynamics approximation, the evolution of such nonhomogeneous medium obeys the equation

$$\frac{\partial \Phi}{\partial t} = [\omega(r) - \omega_0] + a(\nabla\Phi)^2 + b\,\Delta\Phi \;. \tag{4.3.1}$$

After nonlinear transformation (4.1.16) it takes the form

$$\frac{\partial Q}{\partial t} = b\,\Delta Q + \left(\frac{a}{b}\right)[\omega(r) - \omega_0]Q \;, \tag{4.3.2}$$

which is formally equivalent to the quantum Schrödinger equation

$$i\hbar\frac{\partial \Psi}{\partial t} = -\left(\frac{\hbar^2}{2m}\right)\Delta\Psi + U(r)\Psi \tag{4.3.3}$$

if we put

$$\begin{aligned} \Psi &\rightarrow Q \;, \quad \hbar \rightarrow 1 \;, \quad 1/2m \rightarrow b \;, \\ U(r) &\rightarrow -(a/b)[\omega(r) - \omega_0] \end{aligned} \tag{4.3.4}$$

and consider imaginary time $it \rightarrow t$. Note that, in terms of the Schrödinger equation, any region with an increased oscillation frequency $\omega(r)$ corresponds to a certain potential well.

Equation (4.3.2) has a general solution

$$Q(r, t) = \sum_n C_n \exp(\lambda_n t)\mu_n(r) \;, \tag{4.3.5}$$

---

[3] Certainly, this does not exclude existence of autonomous pacemakers in some special reaction-diffusion systems that would be stable with respect to sufficiently small perturbations. *Vasil'ev* and *Polyakova* [4.26] (see also [4.24]) propose a model of a plausibly stable autonomous pacemaker in a three-component reaction-diffusion system, where the inertial third component modifies the local oscillation frequency of the basic activator-inhibitor subsystem and is influenced in turn by the local frequency of these oscillations. Hence, such a third component can create a local inhomogeneity for the oscillating subsystem, which is maintained by the reverse action of the oscillating subsystem on the third component.

where λ_n and $\mu_n(r)$ represent a solution to an eigenvalue problem

$$L\mu = \lambda\mu \tag{4.3.6}$$

for a linear operator

$$L = b\Delta + (a/b)\left[\omega(r) - \omega_0\right] . \tag{4.3.7}$$

Its positive eigenvalues λ_n correspond to bound states of a particle in the potential well $U(r)$. Negative eigenvalues λ form a continuous spectrum. Since $U(r) \to 0$ for $r \to \infty$, there is also a zero eigenvalue $\lambda_0 = 0$ with a corresponding eigenfunction $\mu_0(r)$. Note that the contribution to Q from a continuous spectrum can be neglected because it vanishes exponentially with time.

Wave functions of bound states are localized, i.e. they decrease exponentially at large distances from the well. Thus, utilizing the analogy with the Schrödinger equation, we find

$$\mu_n(r) \approx \text{const} \cdot \exp(-r/r_n) , r \gg r_0 , \tag{4.3.8}$$

where r_n is the localization radius,

$$r_n = \sqrt{b/\lambda_n} . \tag{4.3.9}$$

A wave function corresponding to a zero eigenvalue has a special asymptotic $\mu_0(r) \approx 1$ at large distances r.

For a moment, we assume that a potential well $U(r)$ contains only one bound state, i.e. it has only one positive eigenvalue λ_1 with a corresponding eigenfunction $\mu_1(r)$. Then, neglecting the contribution from the continuous spectrum and taking into account (4.2.4) and (4.2.6), we find

$$\Phi(r, r) = (b/a) \ln\left\{C_0\mu_0(r) + C_1 \exp(\lambda_1/t)\mu_1(r)\right\} . \tag{4.3.10}$$

Far from the region of radius about r_0 where the perturbation of ω is localized, (4.3.10) reduces to

$$\Phi(r, t) = (b/a) \ln\left\{C_0 + C_1 \exp\left[\lambda_1 t - \sqrt{\lambda_1/b}\, r\right]\right\} . \tag{4.3.11}$$

This solution describes a pacemaker. Inside a spherical region of radius $R_p(t) = (b\lambda_1)^{1/2}t$ the second term dominates and we have a system of expanding periodic concentric waves:

$$\Phi(r, t) \approx (b/a) \left[\lambda_1 t - \sqrt{\lambda_1/b}\, r\right] . \tag{4.3.12}$$

These waves are generated at a frequency $\omega = \omega_0 + (b/a)\lambda_1$. Outside this region, uniform oscillations with frequency ω_0 take place. Hence, a pacemaker is localized within a sphere of radius R_p. Since R_p increases with time, such a pacemaker grows until it finally takes over the entire medium.

If a potential well, corresponding to a frequency perturbation $\omega(r)$, has several bound states (i.e. there are several positive eigenvalues λ_n), the analysis is not significantly modified. The only difference is that we should replace λ_1 in (4.3.11) and (4.3.12) by the largest eigenvalue λ_n (which corresponds to the deepest level in this potential well).

In one- and two-dimensional media, even an extremely shallow potential well always has at least one bound state (see [4.27]). In such media, even a very small local rise of the oscillation frequency creates a pacemaker. On the other hand, a potential well in a three-dimensional medium contains a bound state only if it is sufficiently deep. Reformulating the quantum-mechanical condition [4.27] for the existence of a bound state, we find that the linear operator L has at least one positive eigenvalue λ if

$$\delta\omega \gg b^2/ar_0^2 , \tag{4.3.13}$$

where $\delta\omega$ is the characteristic magnitude of the frequency perturbation and r_0 is the characteristic size of the region inside which this perturbation is localized. When a perturbation is so weak that condition (4.3.13) is violated, no pacemaker is created in a three-dimensional oscillatory medium.

We should remember that the phase dynamics equation (4.3.7) is justified only for sufficiently smooth distributions of phase. Specifically, for a pacemaker this implies that the localization radius of an eigenfunction $\mu(r)$, which corresponds to the maximal positive eigenvalue λ_{max}, should be much larger than $(b\tau_{rel})^{1/2}$. Since, according to (4.3.9), this localization radius is $(b/\lambda_{max})^{1/2}$, a condition $\lambda_{max}\tau_{rel} \ll 1$ should hold. Moreover, since inequality $\lambda_{max} \leq (a/b)\delta\omega_{max}$ is always valid, a condition for the applicability of a phase dynamics approximation can be written as

$$\delta\omega_{max} \ll b/a\tau_{rel} . \tag{4.3.14}$$

It was shown above that, when any two periodic waves with different wavenumbers (and different frequencies) collide, they quench one another completely. A narrow layer, within which such quenching occurs, moves towards a region occupied by the wave with a lower frequency. Hence, if we have two different wave generators in the same oscillatory medium, the higher-frequency generator will completely suppress after some time the action of the generator with the lower frequency. Therefore, *competition* between the pacemakers should take place.

If originally there were many different pacemakers, with a random set of frequencies, present in the medium, competition among them will lead to subsequent suppression of pacemakers with increasingly higher frequencies, until a single pacemaker with the maximal available frequency is left. This process was investigated by *Mikhailov* and *Engel* [4.28].

4.4 Spiral Waves

The properties of spiral waves in oscillatory media are in many respects similar to those discussed in Chap. 3 in the case of excitable media. Far from the center, the line of a constant phase has the form of an archimedian spiral. The rotation frequency of a spiral wave is uniquely determined by the parameters of the oscillatory medium, and does not depend on the initial conditions that led to the formation of the wave pattern.

Closer to the center of a spiral wave, the amplitude of oscillations steadily decreases, reaching zero at the central point. Therefore, in oscillatory media we can also use the concept of a spiral wave core, if we interpret it as a region where the amplitude of oscillations is significantly less than in the rest of the medium. However, in contrast to the case of excitable media, the size of the core in an oscillatory medium is usually small, of the same order of magnitude as the diffusional length $(b\tau_{\rm rel})^{1/2}$.

Note that, since within the core the oscillation amplitude undergoes essential changes on a diffusional length scale, the phase dynamics equation (4.1.2) is violated there. Therefore, we cannot construct a closed description of spiral waves (and in particular we cannot determine their rotation periods) by using the approximation of the phase dynamics. This solution can be found only if we return to the original complete equations, with partial derivatives, that describe a given active medium. Hence, the properties of spiral waves in oscillatory media are more intimately related to the particular form of the equations.

Early mathematical studies of spiral waves in oscillatory media were carried out by *Cohen* et al. [4.29], *Erneux* and *Herchkowitz-Kaufman* [4.30], *Greenberg* 4.31, 32], *Mikhailov* and *Uporov* [4.10], *Zeldovich* and *Malomed* [4.11, 4.33] and others. However, the results of these papers were of limited applicability because they did not give any recipes to determine the rotation frequency of a spiral wave, which constitutes its basic property. A consistent mathematical theory of spiral waves in quasi-harmonic oscillatory media was proposed by *Hagan* [4.34]. It describes spiral waves in a $\lambda - \omega$ model (4.1.4), a special case of which is the Ginzburg-Landau equation (4.1.14). Below we follow essentially the analysis by Hagan.

First let us examine the case when $D_2 = 0$, i. e. when the imaginary correction to the effective diffusion coefficient is absent and $D_1 = D$. Note that uniform self-oscillations in this medium have an amplitude ϱ_0 determined by the condition $\lambda(\varrho_0) = 0$ and their frequency is $\omega_0 = \omega(\varrho_0)$. Here, the diffusional length is

$$L_{\rm dif} = \sqrt{b\tau_{\rm rel}} = \sqrt{D/\varrho_0 |\lambda'(\varrho_0)|} \ . \tag{4.4.1}$$

If we introduce a polar coordinate system (r, φ), the spiral wave rotating with frequency ω_* is described by a solution of the form

$$\varrho = \varrho(r) \, , \quad \varPhi = \varphi + \chi(r) + (\omega_* - \omega_0)t \ . \tag{4.4.2}$$

---

[4] In the special case of media where the nonlinear shift of the oscillation frequency is absent, the rotation period of spirals coincides with that of uniform oscillations. However, then the "spiral wave" degenerates into a rotating straight ray.

Substitution of expressions (4.4.2) into (4.1.9) and (4.1.10) with $D_2 = 0$ and $D_1 = D$ gives

$$\varrho_{rr} + (1/r)\varrho_r + \left(D^{-1}\lambda(\varrho) - r^{-2} - \chi_r^2\right)\varrho = 0 , \tag{4.4.3}$$

$$\chi_{rr} + (1/r)\chi_r + 2(\varrho_r/\varrho)\chi_r = (\omega_* - \omega(\varrho))/D . \tag{4.4.4}$$

If we require that $\varrho(r)$ and $\chi_r(r)$ remain bounded at $r = 0$, this further implies

$$\varrho(0) = 0 , \quad \chi_r(0) = 0 . \tag{4.4.5}$$

We should also require that $\varrho(r)$ tends to some nonzero value in the limit $r \to \infty$. Indeed, the amplitude of oscillations cannot increase without any bound. On the other hand, it cannot vanish, either, as $r \to \infty$, because the rest state is unstable in the $\lambda - \omega$ model with respect to small perturbations. Finally, by more subtle arguments (see [4.34]) it can be shown that any solution with an oscillating function $\varrho(r)$ at large r is also unstable.

Suppose that $\varrho \to \varrho_*$ as $r \to \infty$. Then it follows from (4.4.3) and (4.4.4) that

$$\omega_* = \omega(\varrho_*) , \quad \chi_r(r) \to \pm k_* \text{ at } r \to \infty \tag{4.4.6}$$

where $k_* = \left[\lambda(\varrho_*)\right]^{1/2}$.

Therefore, at large distances from the center, such a spiral has a constant step $h_0 = 2\pi/|\chi_r| = 2\pi/k_*$, i.e. it is archimedian. This step is equal to the spatial period of a plane wave with frequency ω_*.

It can be shown that conditions (4.4.5) and (4.4.6) are compatible only for one, uniquely defined, choice of the unknown parameter ω_* in (4.4.3) and (4.4.4). This determines the rotation frequency of spiral waves.

Since differential equations (4.4.3) and (4.4.4) do not permit a complete analytical solution, we are compelled to rely on some further approximations.

First, let us construct, following *Greenberg* [4.31], the solution in a special case when $\omega(\varrho) = \omega_0 = \text{const}$, i.e. when the nonlinear shift of the oscillation frequency is absent. In this case interactions between different elements of the medium do not affect the frequency of their oscillations. Obviously, the rotation frequency of a spiral wave is then ω_0 too. Furthermore, (4.4.4) implies in this case that, if $\chi_r(0) = 0$, we have $\chi_r(r) = 0$ for all r. But this means that $\chi(r)$ is a constant, and according to (4.4.2) such a spiral degenerates into a straigth ray rotating at an angular velocity ω_0.

In this special case the oscillation amplitude $\varrho(r)$ obeys the equation

$$\varrho_{rr} + (1/r)\varrho_r - (\varrho/r^2) + \varrho\lambda(\varrho)/D = 0 , \tag{4.4.7}$$

which should be supplemented by the boundary conditions $\varrho(0) = 0$ and $\varrho(r) \to \text{const}$ as $r \to \infty$. Note that then the asymptotic oscillation amplitude $\varrho(\infty)$ coincides with the amplitude ϱ_0 of uniform oscillations.

If we know function $\lambda(\varrho)$, equation (4.4.7) with such boundary conditions can be numerically integrated, leading to some dependence $\varrho = P_0(r)$ of the oscillation amplitude on r. Figure 4.1 shows this function $P_0(r)$, found by a numeric integration for a medium described by the generalized Ginzburg-Landau equation (4.1.14).

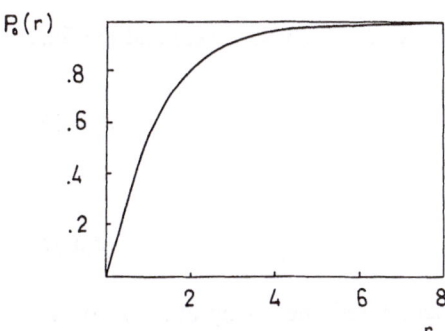

Fig. 4.1. Plot of the function $P_0(r)$

Let us consider further a medium with a nonlinear frequency shift where, however, dependence of the frequency on the oscillation amplitude is very weak, i.e.

$$\omega(\varrho) = \bar{\omega} + \varepsilon\delta\omega(\varrho),\ 0 < \varepsilon \ll 1 .\tag{4.4.8}$$

We introduce a small parameter q, defining it as

$$q = \frac{\omega'(\varrho_0)}{|\lambda'(\varrho_0)|} = \varepsilon\frac{\delta\omega'(\varrho_0)}{|\lambda'(\varrho_0)|},\ |q| \ll 1 .\tag{4.4.9}$$

The solution for spiral waves in such medium can be found by a *method of matched asymptotic expansions*. First, we construct the approximate solutions of (4.4.3) and (4.4.4) in the *inner* $[r \ll L_{dif}\exp(1/|q|)]$ and the *outer* $(r \gg L_{dif})$ regions. Since $|q| \ll 1$, these two regions overlap. Matching of two approximate solutions within their overlap interval allows us to find the unknown parameter ω_* that enters into (4.4.4), i.e. the rotation frequency of spiral waves.

Omitting the details of the calculation, which can be found in [4.34], we present the final results.

The rotation frequency ω_* of spiral waves and the wavenumber k_* are

$$\omega_* = \omega_0 - qDk_*^2 ,\tag{4.4.10}$$

$$k_* = \kappa_*(|q|L_{dif})^{-1}\exp(-\pi/2|q|) .\tag{4.4.11}$$

The numeric factor κ_* depends on the particular form of functions $\lambda(\varrho)$ and $\omega(\varrho)$; for the generalized Ginzburg-Landau equation $\kappa_* = 0.720....$

In the inner region we have the following expansion in powers of q:

$$\varrho(r) = P_0(r) + q^2 P_1(r) + \ldots ,$$
$$\chi_r(r) = |q|v_0(r) + |q|^3 v_1(r) + \ldots .\tag{4.4.12}$$

For instance,

$$v_0(r) = -\frac{\lambda'(\varrho_0)}{\delta\omega'(\varrho_0)}[rP_0(r)^2]^{-1}\int_0^r y_0(y)^2\left[\varepsilon\delta\omega(\varrho_0) - \delta\omega(P_0(y))\right]dy .\tag{4.4.13}$$

If $L_{dif} \ll r \ll L_{dif}\exp(1/|q|)$, we have

$$v_0(r) = (L_{dif}/r)\left[\ln(r/L_{dif}) + C\right] .\tag{4.4.14}$$

The outer region can be divided into the middle ($L_{dif} \ll r \ll 1/k_*$) and the final ($r \gg 1/k_*$) regions. In the middle region

$$\chi_r(r) \approx (L_{dif}/r) \tan \left\{ |q| \left[\ln(r/L_{dif}) + C \right] \right\} \, . \tag{4.4.15}$$

In the final region we find

$$\chi_r(r) \approx -k_* \frac{K_0'(|q|kr)}{K_0(|q|kr)} \, , \tag{4.4.16}$$

where $K_0(x)$ is a zero-order modified Bessel function with the asymptotic $K_0(x) \to 0$ for $x \to \infty$. In particular, at $qkr \ll 1$ the function χ_r is given by the expansion

$$\chi_r(r) \approx k_* \left\{ 1 + \frac{1}{2|q|kr} - \frac{1}{8(qkr)^2} + \ldots \right\} \, . \tag{4.4.17}$$

In the entire outer region the oscillation amplitude $\varrho(r)$ is related to $\chi_r(r)$ by the equation

$$\varrho(r) = \varrho_0 - \left(\frac{1}{r^2} + \chi_r^2 \right) L_{dif}^2 \, . \tag{4.4.18}$$

When $r \to \infty$, the oscillation amplitude tends to the limit

$$\varrho_* = \varrho_0 - (L_{dif} k_*)^2 \, . \tag{4.4.19}$$

Until now we have considered only oscillatory media where the effective diffusion coefficient D is real. The above results can, however, be generalized to the case when $D = D_1 + D_2$. The only difference is that now, as a small parameter q, the quantity

$$q = \frac{D_1 \left[\omega'(\varrho_0)/|\lambda'(\varrho_0)| \right] - D_2}{D_1 + \left[\omega'(\varrho_0)/|\lambda'(\varrho_0)| \right] D_2} \tag{4.4.20}$$

should be taken; the effective diffusion length thus becomes

$$L_{dif} = \frac{1}{\sqrt{|\lambda'(\varrho_0)|}} \sqrt{D_1 + \frac{\omega'(\varrho_0)}{|\lambda'(\varrho_0)|} D_2} \, . \tag{4.4.21}$$

Note that now the smallness of the nonlinear frequency shift is not sufficient to guarantee the applicability of the matched asymptotic expansions approximation. Indeed, this condition cannot ensure that the parameter q is small (see (4.4.20)).

Remarkably, the parameter q defined by (4.4.20) can be expressed, in a very simple way, in terms of the coefficients a and b entering into the phase dynamics equation (4.1.2), i.e. we have

$$q = -a/b \, . \tag{4.4.22}$$

By taking into account that $L_{dif} = (b\tau_{rel})^{1/2}$ we can rewrite (4.4.11) as

$$k_* = \kappa_* \sqrt{b/a^2 \tau_{rel}} \, \exp(-\pi b/2|a|) \, , \tag{4.4.23}$$

where κ_* is a dimensionless numeric factor.

The rotation frequency ω_* of spiral waves is given by

$$\omega_* = \omega_0 + ak_*^2 . \tag{4.4.24}$$

Substituting k_* from (4.4.23), we see that, although the rotation frequency depends on the particular form of the equations of an oscillatory active medium, this dependence remains rather weak when $|a| \ll b$, since the particular form of the equations is reflected only in the numeric value of the parameter κ_* in (4.4.23). Because of this universality, one can conjecture that expressions (4.4.23) and (4.4.24), derived in the special case of quasi-harmonic self-oscillations, would hold also for an arbitrary oscillatory active medium in the limit $|a| \ll b$, with a different value of the factor κ_*.

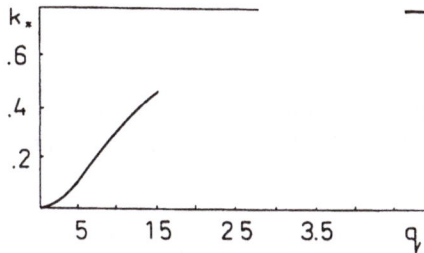

Fig. 4.2. Asymptotic wavenumber k_* as a function of q for one-armed spiral waves in the Ginzburg-Landau equation (from [4.34])

When parameter $|q|$ increases, the wavenumber k_* of the generated waves increases too, approaching values of about L_{dif}^{-1}. In this situation, the rotation frequency ω_* and the wavenumber k_* depend intimately on the particular form of equations for a given oscillatory medium, and the above approximation of matched asymptotic expansions is no longer applicable. The only possibility consists now in numeric integration of (4.4.3) and (4.4.4) with boundary conditions (4.4.5) and (4.4.6). The dependence of k_* on q, computed in [4.34] for the Ginzburg-Landau equation (4.1.14), is shown in Fig. 4.2.

It should be remarked that, for sufficiently large q, spiral waves should lose their stability. Indeed, far from the rotation center their wave pattern looks locally like a plane periodic wave with the wavenumber k_*. When q grows this wavenumber increases. However, plane waves with sufficiently large wavenumbers are unstable; usually the stability is lost at $k \sim 1/L_{\mathrm{dif}}$. Consequently, spiral waves can be found only in such oscillatory media where the parameter $q = -a/b$ is not very large. For example, plane waves in the medium described by the Ginzburg-Landau equation become unstable when

$$kL_{\mathrm{dif}} > (3 + 2q^2)^{-1/2} . \tag{4.4.25}$$

Thus, spiral waves in such medium exist only if $q < 1.397\ldots$.

It is possible to construct solutions for multi-armed rotating spirals. However, the analysis carried out by *Hagan* [4.34] reveals that all such multi-armed patterns are unstable. A two-armed spiral wave decays into two single-armed spiral waves, etc.

Above we considered properties of spiral waves in $\lambda - \omega$ models. However, these wave patterns exist also in other oscillatory media. In this case there is no general theory available (few further applications of the Hagan method are given in [4.35]). Nevertheless, some conclusions concerning the properties of spiral waves can be drawn from the phase dynamics approximation, as indicated by *Kuramoto* (see [4.2]) and *Koga* [4.37, 38].

Although this approximation is violated near the center of a spiral wave, we can still use it at the periphery if the wavenumber k_* of emitted waves is sufficiently small. The form of a steadily rotating spiral wave in the peripheral region can be found from the phase dynamics equation (4.1.2) by substitution of $\Phi = \varphi + \chi(r) + (\omega_* - \omega_0)t$, which gives

$$\omega_* - \omega_0 = b(1/r)\chi_r + b\chi_{rr} + a(\chi_r^2 + 1/r^2) . \tag{4.4.26}$$

If $a = 0$, the solution to (4.4.26) is

$$\chi_r = (1/2b)(\omega_* - \omega_0)r + C/r , \tag{4.4.27}$$

where C is an integration constant. We see that χ_r remains bounded in the limit $r \to \infty$ only if $\omega_* = \omega_0$, i.e. the rotation frequency of spiral waves coincides with ω_0. By further integration we find then

$$\chi = C \ln(r/r_a) , \tag{4.4.28}$$

where r_a is another integration constant. Thus, in oscillatory media with $a = 0$ all spiral waves are *logarithmic*. Note that a special case of a logarithmic spiral is a straight ray, which corresponds to $C = 0$ in (4.4.28).

When $a \neq 0$, equation (4.4.26) reduces for $r \to \infty$ to

$$\omega_* - \omega_0 = b\chi_{rr} + a\chi_r^2 . \tag{4.4.29}$$

Since we require that χ_r should remain bounded in this limit, this implies that $\chi_{rr} \to 0$. Therefore, asymptotically we have

$$\omega_* = \omega_0 + a\chi_r^2 . \tag{4.4.30}$$

Hence, in media with $a > 0$ spiral waves have rotation frequencies larger than the frequency ω_0 of uniform oscillations, whereas for $a < 0$ the inverse relationship holds. The precise value of the rotation frequency ω_* cannot be found within the phase dynamics approximation. Since χ_r has a finite limit as $r \to \infty$, the spiral is archimedian.

Note that in the special case $a = 0$, the phase dynamics equation (4.1.2) becomes linear

$$\frac{\partial \Phi}{\partial t} = b\Delta\Phi \tag{4.4.31}$$

and can be written in a variational form

$$\frac{\partial \Phi}{\partial t} = \frac{\delta F[\Phi]}{\delta \Phi(r, t)} \tag{4.4.32}$$

where

$$F = \int \frac{1}{2} b(\nabla \Phi)^2 d\boldsymbol{r} \ . \tag{4.4.33}$$

Therefore, stable solutions should correspond to minima of the functionals $F[\Phi]$.

When $a = 0$, spiral waves are logarithmic and, to a good approximation, can be represented by straight rotating rays. A ray is described in polar coordinates by an equation $\Phi = m\varphi$ (where $m = \pm 1$, depending on the rotation direction). For such a special solution we have

$$F_1 = \pi b \int_0^\infty \frac{1}{r} dr \ . \tag{4.4.34}$$

This integral diverges. However, we should remember that the phase dynamics equation becomes invalid near $r = 0$, and therefore we cannot integrate at very small r. Moreover, the upper bound of integration is also limited, by a dimension R of the medium. Hence,

$$F_1 \approx \pi b \ln(R/R_0) \tag{4.4.35}$$

where R_0 is about the size of a core region (because of the weak logarithmic dependence, an exact value of R_0 is not important).

Since equation (4.4.31) is linear, a superposition of any two solutions will again be a solution. For instance, in cartesian coordinates two "spiral waves" (rotating rays) with centers located at points (x_1, y_1) and (x_2, y_2) are described by the equation (note that $\varphi = \arctan(y/x)$) :

$$\Phi = m_1 \arctan\left[\frac{(y - y_1)}{(x - x_1)}\right] + m_2 \arctan\left[\frac{(y - y_2)}{(x - x_2)}\right] \ . \tag{4.4.36}$$

Substitution of (4.4.36) into (4.4.33) yields the value of the functional F for these two interacting special spiral waves

$$F_{12} = \pi b(m_1^2 + m_2^2) \ln(R/R_0) - m_1 m_2 \ln(R_{12}/R_0) \tag{4.4.37}$$

where

$$R_{12} = \sqrt{(x_1 - x_2)^2 + (y_1 - y_2)^2} \tag{4.4.38}$$

is the distance between their centers.

Thus, if both rays rotate in the same direction (i. e. $m_1 = m_2$), F becomes smaller for larger distances R_{12}. Since in the process of evolution F can only decrease in time, this implies that such spiral waves should repel one another. On the other hand, when two spiral waves rotate in opposite directions ($m_1 = -m_2$), they are attracted.

An m-armed spiral wave (which corresponds in our case to m rays rotating around the same center) is described by the equation $\Phi = m\varphi$. For such a solution, functional F takes the value

$$F_m \approx \pi b m^2 \ln(R/R_0) \tag{4.4.39}$$

which is m^2 times greater than F_1. At the same time, the total value of the functional F for m one-armed spirals separated by very large distances is equal to mF_1. Therefore, multi-armed spiral waves are unstable since they can decay into a sum of simple spirals, thereby decreasing the value of F.

The above results were derived for a special case when $a = 0$. When a is sufficiently small but nonvanishing, the step of a spiral is very large. If the distance between centers of two spirals remains less than this step, interacting spirals can be well approximated by straight rays. Hence, we can expect that at such distances any two spirals should repel if they rotate in the same direction and attract if their rotation directions are opposite.

Recently, *Aranson* et al. [4.39] have constructed a theory of spiral wave interactions for the generalized Ginzburg-Landau model and some related equations, which uses a similar approach. Note that some estimates for the interaction between spiral waves were given already in [4.11]. *Aranson* et al. [4.39] also discuss migration of spiral wave centers in smoothly inhomogeneous oscillatory media.

Extensive numerical simulations of spiral waves in a generalized Ginzburg-Landau model were carried out by *Aranson* and *Rabinovich* [4.40] and by *Bodenschatz* et al. [4.41]. They confirmed the predictions of the Hagan theory. *Bodenschatz* et al. [4.41] also studied the evolution of large populations of spiral waves; it was found that generally their centers tend to order themselves into a regular lattice. *Elphick* and *Meron* [4.42] analytically demonstrated the existence of bound pairs of spiral waves in the generalized Ginzburg-Landau model.

5. Active Media with Long-Range Inhibition

In this chapter we consider stationary patterns that are formed in a special class of two-component media, where one of the components can be treated as an activator and the other as an inhibitor. Generally, an isolated element of such a medium can be either bistable, or excitable, or oscillatory. This distinction becomes, however, of minor importance if inhibition is long-range (for instance, when the diffusion constant of an inhibitor component is very large) and if it quickly adjusts to the momentary distribution of an activator. These conditions favor the formation of stable stationary dissipative patterns.

There are two basic kinds of such patterns, namely spikes and broad strata. An individual spike represents a small region with an increased density of the activator component which is immersed in a much larger region with increased inhibitor density. Depending on the properties of a particular medium, one can observe either solitary spikes, or sparsely dispersed groups of them, or spike lattices.

When the inhibitor component becomes more inertial, the stability of the spikes decreases, and at some point they start to pulsate.

Broad strata represent large domains, inside which an active medium approaches two different stationary uniform states. Such domains are separated by narrow interfaces (or kink regions) where the activator density is sharply changed.

5.1 Integral Negative Feedback

In addition to propagating waves, distributed active media can maintain various *stationary dissipative patterns*, stable with respect to small perturbations.

In effect, we have already seen an example of a stationary dissipative pattern in Chap. 2 where we studied one-component bistable media. As pointed out in Sect. 2.3, if condition (2.3.8) is satisfied, an interface between the regions occupied by two different steady states ("phases") does not move. Under this condition, a medium is divided into a mixture of domains of the two phases.

However, this example has some peculiarities. In this case the domains remain arbitrary and their spatial form is not recovered after a perturbation. Hence, the system is in a state of neutral (or marginal) equilibrium. Furthermore, condition (2.3.8) can be satisfied only for a special choice of the parameters of a bistable medium.

Generally, an interface in a simple one-component bistable medium moves (and thus represents a propagating trigger wave). The direction and the velocity of its motion depend on the values of the medium parameters. By changing them, one can control the motion of interfaces.

A stable dissipative pattern can be obtained in a most simple way, if we assume that the parameters of a one-component bistable medium are not fixed; instead they quickly adjust to the momentary positions of interfaces. A good illustration of this is provided by the *barretter effect* studied by *Barelko* et al. [5.1].

Let us consider a thin iron wire in a hydrogen atmosphere. The wire is heated by an electric current I that passes through it. It turns out that the specific resistance R of an iron wire immersed in hydrogen depends nonlinearly on its temperature. As we see from Fig. 5.1, function $R = R(\theta)$ is sharply increasing when the temperature exceeds a certain threshold value. The Joule heat production $Q_+ = I^2 R(\theta)$ is counteracted by the heat flow $Q_- = \gamma_0(\theta - \theta_0)$ to the gas at temperature θ_0. Under these conditions, the equation of thermal balance for the wire can be written as

$$C\frac{\partial \theta}{\partial t} = f(\theta) + \chi \frac{\partial^2 \theta}{\partial x^2} , \tag{5.1.1}$$

where $f(\theta) = Q_+ - Q_-$, C is the heat capacity per unit length of the wire, and χ is its thermal conductivity.

Let us assume that current I is kept constant. Then (5.1.1) coincides with a general equation (2.1.13) of one-component bistable media. Within the interval $I_+ > I > I_-$ this system possesses two stable steady states (Fig. 5.2), one of them with a high temperature of the wire ($\theta = \theta_3$) and another with a low temperature ($\theta = \theta_1$). These states can be considered as two different "phases". In a distributed system described by (5.1.1), transitions between the states can be realized by propagation of a temperature front (i.e., of a trigger wave) that moves into a high-temperature or a low-temperature region, depending on the parameters of the system. As shown in Chap. 2, the direction of motion of a trigger wave is determined by the sign of the integral

$$A = \int_{\theta_1}^{\theta_3} f(\theta) \, d\theta . \tag{5.1.2}$$

If $A < 0$, the front propagates into the region occupied by a high-temperature regime $\theta = \theta_3$, whereas at $A > 0$ the low-temperature region is ousted. At $A = 0$, trigger waves change their direction of propagation, so that their velocity vanishes. Then a stationary heterogeneous state of this system becomes possible, i.e. hot and cold strata can coexist on the wire.

Until now we have assumed that the wire is connected to a controlled source of electric current which maintains the current I fixed. When a controlled voltage source is used, which maintains a fixed voltage difference U on the ends of the wire, the observed phenomena are different. In this case current I varies with the total resistance of the wire as

$$I = U \left[\int_{-L_0}^{L_0} R(\theta(x)) \, dx \right]^{-1} , \tag{5.1.3}$$

where $2L_0$ is the wire length.

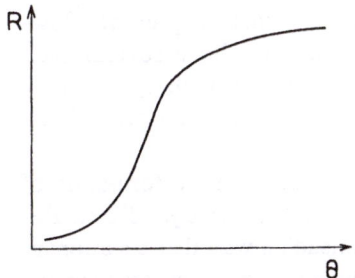

Fig. 5.1. Specific resistance R of an iron wire in the hydrogen atmosphere versus temperature θ

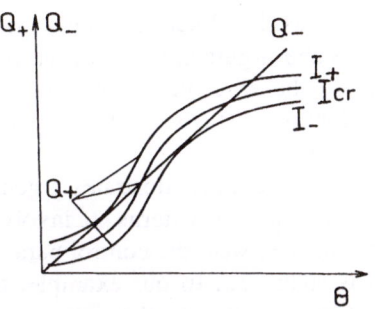

Fig. 5.2. Heat production Q_+ for three different values of current I and heat flow Q_- versus temperature θ. The intersection points give the steady states of the system

Suppose that current I passing through the cold wire (with $\theta = \theta_1$) has exceeded the critical magnitude I_{cr} at which $A = 0$. Then if we introduce a local perturbation (i. e. heat up the wire locally), it will initiate two propagating fronts of transition into a high-temperature state $\theta = \theta_3$ that produce a growing high-temperature domain. Since the high-temperature region will grow, the total resistance of the wire will increase. This will lead to a decrease of the current I, and hence of the quantity A. When I drops down to I_{cr}, the fronts will stop. Thereby, the wire will become divided into a hot domain (with a high temperature θ_3 and a high resistance $R(\theta_3)$ and the remaining cold region with a small resistance $R(\theta_1)$. If we increase the voltage difference U, this will change the length of the hot domain, and consequently the total resistance of the wire. However, this will *not* influence the current passing through the wire; it is always equal to I_{cr} so that the cold and the hot regions can steadily coexist. Therefore, such a system operates as a current stabilizer, keeping the current fixed at I_{cr} so long as a high-temperature domain does not spread over the entire wire. A similar process takes place when U is decreased. This effect is utilized in a technical device known as a *barretter*.

Suppose that the width $2l_3$ of a high-temperature domain is much larger than the width $2l_0$ of an interface layer. Then we can approximately write down the current passing through the wire as

$$I = \frac{U}{2(R_3 l_3 + R_1 l_1)} , \quad l_1 = L_0 - l_3 , \tag{5.1.4}$$

where $R_3 = R(\theta_3)$ and $R_1 = R(\theta_1)$. Taking into account that in the established regime the critical current I_{cr} is maintained, we find from (5.1.4) that

$$l_3 = (R_3 - R_1)^{-1} \left[(U/2I_{cr}) - R_1 L_0 \right] . \tag{5.1.5}$$

Therefore, a stationary dissipative pattern, namely the hot domain, first appears at $U = U_{min}$, where $U_{min} = 2L_0 R_1 I_{cr}$. When U further increases, this domain grows and at $U = U_{max}$ (where $U_{max} = 2L_0 R_3 I_{cr}$) it spreads over the entire wire. For $U > U_{max}$, only a uniform high-temperature state is stable.

We have assumed here that only one high-temperature domain is present. However, the same arguments can be used in a situation when we have several such domains on the wire, provided that their total length coincides with l_3 given by (5.1.5). How many such hot domains exist and where they are located depends on the initial conditions.

The above example illustrates a general mechanism leading to the formation of stationary dissipative patterns. It involves *quick long-range negative feedback*. Because of this feedback, the control parameter changes in such a way that a stationary domain is stabilized. In our example, the feedback, which is realized through the current I, is sensitive to the momentary states of *all* elements of the distributed system. Hence, it has an *integral* character.

Evidently, other processes leading to a similar effect are possible. Suppose that some fuel is burned in a medium. The diffusion constant of this substance is much larger than thermal conductivity of the medium. Under this condition, the fuel concentration s is maintained practically constant everywhere in the medium due to fast diffusion. Furthermore, we assume that fuel is continuously supplied into the medium at a rate W. These processes are described by the equations

$$\partial\theta/\partial t = sq(\theta) - \gamma_0(\theta - \theta_0) + \chi\Delta\theta$$

$$ds/dt = W - s \int p\big(\theta(r)\big)\, dr . \tag{5.1.6}$$

Here $q(\theta)$ is the specific heat production rate, and $p(\theta)$ is the specific rate at which the fuel is spent in the process of combustion. Since combustion begins only when a certain threshold temperature is exceeded, both these functions have the form of a smoothed step (see Fig. 2.2).

Equations (5.1.6) describe the formation of a stationary dissipative pattern. Suppose we have ignited the fuel inside a certain small region of this medium. Then the combustion seat will start to spread out. However, the spreading of this zone is accompanied by an increase in consumption of the fuel, and consequently leads to a decline in the fuel concentration s. In its turn, a decline in s results in slowing down of the combustion front. When a certain critical concentration a_{cr} is reached, the front stops. By this process, a localized combustion seat is formed in the medium, which can be considered as another example of a stationary dissipative pattern.

Here we again have *multiplicity* of dissipative patterns. Instead of a single large combustion seat there might be several smaller seats with the same total consumption of fuel. The positions and total number of such seats are determined by the initial conditions. If the combustian seats are spaced not very far one from another compared to the width l_0 of the combustion front, their interaction can be observed: neighboring seats are attracted and tend to fuse together.

Note that our arguments have used the assumption that the fuel is spent very rapidly compared with the characteristic rates of temperature variation. Therefore, the fuel concentration s can adjust adiabatically to the momentary distribution of temperature. When the fuel dynamics is slower, such adjustment occurs with some delay. When this delay exceeds a certain critical value, the stationary regime becomes unstable and undamped pulsations of the combustion zone may appear. This can be interpreted as the formation of an *oscillatory dissipative pattern*.

A long-range feedback can also be realized by means of a rapidly diffusing inhibitor. In this case, unlimited growth of the combustion seat is prevented by an increase in the inhibitor concentration.

5.2 Spike Patterns

The introductory examples of Sect. 5.1 illustrate the basic mechanisms which lead to the formation of stable stationary dissipative patterns. Now we can examine more formal aspects of the mathematical theory of such patterns in two-component reaction-diffusion media.

Stationary dissipative patterns were first studied by *Turing* [5.2] in a model for biological pattern formation. Later *Prigogine* and *Lefever* [5.3, 4] proposed a hypothetic chemical reaction model (the *Brusselator*) which was used for extensive studies of such patterns. At present, there is much literature dealing with different aspects of the formation, stability and properties of stationary dissipative patterns in various active media (see the books by *Nicolis* and *Prigogine* [5.5], and *Vasilev* et al. [5.6] and the review by *Belintsev* [5.7]). In this chapter we follow mainly the approach by *Kerner* and *Osipov* [5.8-12], which is based on the methods of singular perturbation theory.

We consider a general model which describes the interaction of a slowly diffusing activator u with a rapidly diffusing inhibitor v. Unless it is explicitly specified we assume that the dynamics of the inhibitor is very swift in comparison to the characteristic rates of change of the activator concentration. This model is given by the equations

$$\tau_u \dot{u} = f(u, v) + l^2 \Delta u$$
$$\tau_v \dot{v} = g(u, v) + L^2 \Delta v \ . \tag{5.2.1}$$

Here l and L ($L \gg l$) are the characteristic diffusion lengths of the activator and the inhibitor, respectively. Parameters τ_u and τ_v represent the characteristic times of the activator and the inhibitor.

We assume that production of the inhibitor v increases with the activator concentration, i.e. g is a monotonously increasing function of u. On the other hand, at a fixed activator concentration, the growth of v is saturated because g declines with increasing v. The simplest function which satisfies these two assumptions is linear,

$$g(u, v) = au - bv \ , \tag{5.2.2}$$

with positive coefficients a and b.

At a fixed concentration of the inhibitor, function f has the form shown in Fig. 2.1; within a certain range of values of u it has a positive derivative $\partial f / \partial u$. This is effectively equivalent to the autocatalytic reproduction of the activator u. Since v *inhibits* reproduction of u, we assume that the derivative $\partial f / \partial v$ is negative. In the simplest case we can write

$$f(u, v) = f(u) - cv \ , \tag{5.2.3}$$

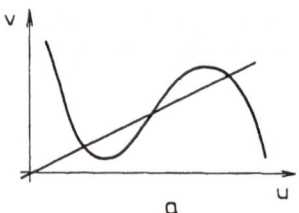

Fig. 5.3a–c. The null-clines of isolated active elements: **(a)** a bistable element, **(b)** an excitable element, **(c)** an oscillatory element

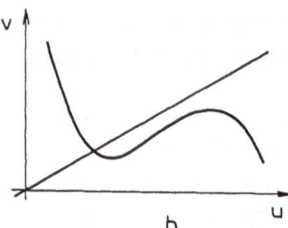

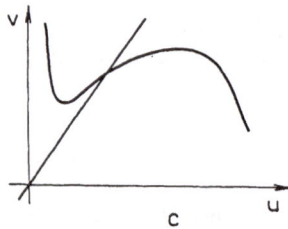

where $f(u)$ is some function with the form given by Fig. 2.1 and c is a positive parameter.

An isolated element of the active medium, described by (5.2.1), has the null-clines $f(u, v) = 0$ and $g(u, v) = 0$ shown in Fig. 5.3. Depending on the relative position of these two null-clines, it can be either bistable (Fig. 5.3a), or excitable (Fig. 5.3b), or oscillatory (Fig. 5.3c). However, including long-range inhibition deeply modifies the behavior of the interacting elements. As a result, the dissipative patterns in media composed of different active elements turn out to be very similar.

Basically, they can be divided into two large classes, i.e. *spike patterns* and *broad strata* (which will be discussed in the next section). An individual spike usually represents a small spot with high activator concentration which is immersed in a larger region with increased concentration of the inhibitor.

From a general point of view, the prototype of a solitary spike is the *critical nucleus* in a one-component bistable medium which was described in Sect. 2.3. In the latter case, however, such a nucleus is unstable: when a small perturbation is applied it either shrinks (and disappears) or expands. If an inhibitor is added, it can stabilize such a pattern. Periodic or irregular spike lattices also exist.

Let us consider first a situation when the medium is one-dimensional and its size L_0 is much smaller than the characteristic diffusion length L of the inhibitor (but still much larger than the diffusion length l of the activator). Then the spatial distribution of the inhibitor is uniform and (5.2.1) reduces to equations

$$\tau_u \dot{u} = f(u, v) + l^2 \Delta u$$
$$\tau_v \dot{v} = \frac{1}{L_0} \int_{-L_0/2}^{L_0/2} g(u, v)\, dx \ . \tag{5.2.4}$$

The right side of the second equation is simply the average production of the inhibitor. At the ends of the segment $(-L_0/2, L_0/2)$ we impose the boundary conditions

$$\partial u / \partial x = 0 \ , \tag{5.2.5}$$

which reflect the absence of diffusion flow.

A solitary spike located at the center of the segment is a symmetric stationary solution of (5.2.4). The form of the spike $u = u_0(x)$ is determined by the equation

$$0 = f(u_0, v_0) + l^2 \partial^2 u_0 / \partial x^2 \tag{5.2.6}$$

at a given constant value v_0 of the inhibitor concentration. This equilibrium concentration can be found from the condition

$$\int_{-L_0/2}^{L_0/2} g\big(u_0(x), v_0\big)\, dx = 0 \ , \tag{5.2.7}$$

where $u_0(x)$ is the solution to (5.2.6).

The characteristic width of the spike is of the same order of magnitude as the diffusional activator length l, which is the only length parameter entering into (5.2.6). When $L_0 \gg l$, the influence of the spike does not extend to the borders of the medium. Then it coincides with the solitary critical nucleus in an infinite one-component bistable medium, which was analyzed in Sect. 2.3. This becomes evident if we compare equations (2.3.5) and (5.2.6). In particular, the activator concentration $u^* = u_0(0)$ in the center of the spike is determined by condition (2.3.7), where $f(u)$ should be replaced now by $f(u, v_0)$.

To analyze the stability of a solitary spike we should investigate the evolution of its small perturbations:

$$u(x, t) = u_0(x) + \delta u(x, t) \ ,$$
$$v(t) = v_0 + \delta v(t) \ . \tag{5.2.8}$$

Substituting (5.2.8) into (5.2.4) and performing linearization with respect to small perturbations, we find the equations

$$\tau_u \delta \dot{u} = \hat{\Gamma} \delta u - B(x) \delta v \ ,$$
$$\tau_v \delta \dot{v} = -\gamma \delta v + \frac{1}{L_0} \int_{-L_0/2}^{L_0/2} A(x) \delta u(x)\, dx \ . \tag{5.2.9}$$

Here we have introduced the notations:

$$B(x) = -\partial f/\partial v|_{u_0(x),v_0} \, , \tag{5.2.10}$$

$$A(x) = \partial g/\partial u|_{u_0(x),v_0} \, , \tag{5.2.11}$$

$$\gamma = -\frac{1}{L_0} \int_{-L_0/2}^{L_0/2} (\partial g/\partial v)\,|_{u_0(x),v_0} \, dx \, . \tag{5.2.12}$$

Note that all these quantities are positive because $\partial f/\partial v < 0$, $\partial g/\partial u > 0$, and $\partial g/\partial v < 0$.

The linear differential operator $\hat{\Gamma}$ is defined as

$$\hat{\Gamma} = l^2 \frac{\partial^2}{\partial x^2} + \frac{\partial f}{\partial u}\,|_{u_0(x),v_0} \, . \tag{5.2.13}$$

It is self-adjoint and therefore all its eigenvalues λ_k are real. Note that these eigenvalues determine the evolution of the critical nucleus which is described by the first of the equations (5.2.4) under the condition that the inhibitor concentration v is kept constant. Positive values of λ_k correspond to growing perturbation modes, whereas the negative eigenvalues correspond to damped modes. Since such a critical nucleus is unstable, operator $\hat{\Gamma}$ always has at least one positive eigenvalue.

The eigenfunctions $\Phi_k(x)$ of this linear operator represent a complete set, so that any other function can be decomposed into their linear superposition. Using this property, we write

$$\delta u(x,t) = \sum_k C_k(t)\Phi_k(x) \, , \tag{5.2.14}$$

$$B(x) = \sum_k B_k\Phi_k(x) \, , \quad A(x) = \sum_k A_k\Phi_k(x) \, . \tag{5.2.15}$$

Substitution of (5.2.14) and (5.2.15) into (5.2.9) yields

$$\tau_u \dot{C}_k = \lambda_k C_k - B_k \delta v \, ,$$
$$\tau_v \dot{\delta v} = -\gamma \delta v + \sum_k A_k C_k \, , \tag{5.2.16}$$

We can seek the solution of these linear equations in the form $\delta v \sim C_k \sim \exp(\mu t)$. Then, as can easily be verified, μ satisfies the algebraic equation:

$$\tau_v \mu + \gamma = \sum_k \frac{A_k B_k}{\lambda_k - \tau_u \mu} \, . \tag{5.2.17}$$

The spike is stable with respect to small perturbations if $\mathrm{Re}\{\mu\} < 0$ for any root $\hat{\mu}$ of (5.2.17).

The properties of the roots can be found from the graphic solution of (5.2.17) which is obtained by plotting the left and the right sides of this equation (see (Fig. 5.4).

Every eigenvalue γ_k gives rise to a pole in the dependence of the right side on μ. Examination of the graphic solution reveals (Fig. 5.4) that (5.2.17) always have at least one positive real root μ if there are two or more positive eigenvalues λ_k of the operator $\hat{\Gamma}$. Hence, stability of the spike requires that there is exactly one

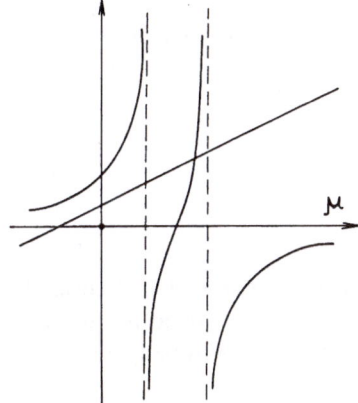

Fig. 5.4. Graphic solution of equation (5.2.17)

positive eigenvalue of this linear differential operator. However, this condition is not yet sufficient. If $\tau_v \to 0$ and all negative eigenvalues have much greater magnitude than the only positive eigenvalue λ_1, the maximal root of equation (5.2.17) can be estimated as

$$\mu_1 = (1/\tau_u)(\lambda_1 - A_1 B_1/\gamma) \, . \tag{5.2.18}$$

Therefore, the spike is stable if condition

$$A_1 B_1 > \lambda_1 \gamma \tag{5.2.19}$$

holds. This implies strong coupling between the activator and the inhibitor variations.

If the characteristic time τ_v of the inhibitor variation is not very small, the largest root μ_1 can become complex. This happens if condition

$$A_1 B_1 > \lambda_1 \gamma + (\tau_u \tau_v)^{-1}(\gamma \tau_u - \lambda_1 \tau_v)^2 \tag{5.2.20}$$

is satisfied (we assumed that $\lambda_1 \gg |\lambda_k|$ for all $k > 1$). The real part of this complex root is positive if

$$\lambda_1 \tau_v > \gamma \tau_u \, . \tag{5.2.21}$$

When this inequality holds, the spike is unstable with respect to oscillations of its width and form.

Within the linear approximation, the amplitude of such oscillations grows exponentially in time. However, nonlinear terms can saturate its growth. As a result, stable periodic *pulsations* of the spike can appear[1]. Such oscillatory dissipative patterns were analyzed by *Kerner* and *Osipov* [5.8, 13-14].

Until now we have assumed that the linear size L_0 of the medium is small compared with the characteristic diffusional length L of the inhibitor (and hence that its distribution is uniform). In larger media a spike consists of a small core with a high activator concentration and a larger region where the inhibitor concentration is increased (see Fig. 5.5).

---

[1] These pulsations might also be chaotic, i.e. they can be described by some strange attractor.

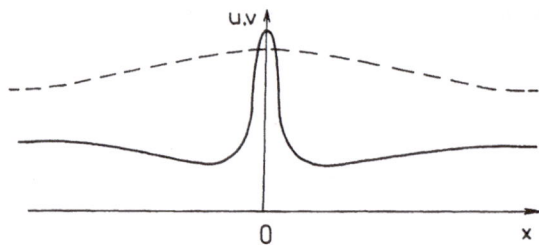

Fig. 5.5. Distributions of the activator (*solid line*) and the inhibitor (*dashed line*) in a spike pattern

The core of the spike represents a stabilized critical nucleus, i. e. it is described by (5.2.6). The inhibitor concentration v_0 in the core is now determined not by the integral solution (5.2.7), but by another consistency condition that will be formulated below.

In the larger surrounding region, activator concentration u adjusts to the local inhibitor concentration v, i. e. $u = u(v)$. The dependence $u(v)$ is given by the solution of equation $f(u, v) = 0$ (namely, by the branch of the solution which corresponds to the lowest activator concentration).

When the characteristic diffusion lengths l and L of the activator and the inhibitor are very different, the core can be modeled in the equation for the inhibitor concentration v by a delta-function source, i. e. we can write this equation in the stationary case as

$$\nu\delta(x) + g\big(u(v), v\big) + L^2 \partial^2 v / \partial x^2 = 0 \,. \tag{5.2.22}$$

Since the delta-function source models additional production of the inhibitor in the small core, ν should coincide with the actual excessive production of the inhibitor in the core region. This condition implies

$$\nu = \int \big[g\big(u_0(x), v_0\big) - g\big(u(v_0), v_0\big)\big] \, dx \tag{5.2.23}$$

where $u_0(x)$ is the activator distribution in the core region that can be found (for any given value of v_0) from (5.2.6); integration is performed over the core region. We should add also boundary conditions $\partial v / \partial x = 0$ at the ends $x = \pm L_0 / 2$ of the segment.

In effect, (5.2.22) determines the complete inhibitor concentration distribution $v(x)$, including its value $v(o)$ in the core center. But $v(0)$ should coincide with the value v_0 of the inhibitor concentration which was assumed when we calculated the activator distribution in the core region. This consistency condition can be used to find v_0, thus finishing the construction of the spike solution.

Note that the same solution also describes a periodic spike pattern, with a spatial period L_0.

Principally, stability of spike patterns with varying inhibitor concentration can be analyzed along the same lines as was done above in the case of extremely fast inhibitor diffusion. However, the analysis becomes much more complicated because we should consider nonuniform perturbations of the inhibitor distribution. This investigation was carried out by *Kerner* and *Osipov* [5.8, 9, 12]. They have found that there is a minimal spatial period $L_{\min}$ of stable spike patterns. If we produce, as an initial condition, a more frequent sequence of spikes, it will undergo restructuring: some of the spikes will merge with their neighbors and the others rearrange, so that finally we find a pattern with a spatial period not exceeding L_{min}.

In some cases there are also upper limitations on the spatial period L_0 of stable spike patterns.

If $L_0 \rightarrow \infty$ we have a solitary spike in an infinite medium. Obviously, such a solution is permitted only if our active medium has (at least one) stable uniform stationary state, i.e. if its individual elements are either bistable or excitable (Fig. 5.3a,b). For such media the spatial period of stable spike patterns is not limited from above. On the other hand, in active media which are composed of oscillatory elements (Fig. 5.3c) there is no stable uniform stationary state, and hence solitary spikes are impossible. The maximal spatial period L_{max} of spikes in such media is determined by the condition that the inhibitor concentration between the spikes cannot fall lower than its minimal value v_{min} which corresponds to the bottom of the left branch of the null-cline $f(u, v) = 0$ (see Fig. 5.3c).

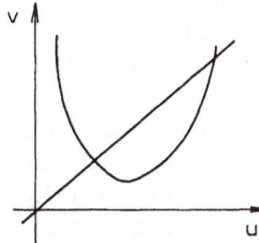

Fig. 5.6. Null-clines of an unstable element

It is interesting to note that such steady spike patterns can exist even in active media which consist of unstable individual elements. Consider, for instance, an element with the null-clines $f(u, v) = 0$ and $g(u, v) = 0$ shown in Fig. 5.6. Taken apart, this element is unstable: when the initial concentration of the activator exceeds a certain threshold, it starts to grow without any limit. If we fix the inhibitor concentration v and consider a one-component active medium which is composed of such elements and described by the first of equations (5.2.1), we can construct a "critical nucleus", i.e. a stationary localized solution which is absolutely unstable under small perturbations. When it is slightly squeezed, this nucleus shrinks and disappears. On the other hand, if we slightly expand it, the nucleus begins to explode indefinitely. Introduction of a rapidly diffusing strong activator, whose distribution quickly adjusts to the momentary distribution of the activator, can stabilize such a critical nucleus and transform it into a stationary stable spike. Note that the above mathematical treatment is equally applicable to describe spike patterns in such active media.

Obviously, the terms "activator" and "inhibitor" are only useful conventions. In reality they can correspond to a large variety of quantities of very different nature. It is also possible that an active medium is described by equations (5.2.1) but the functions $f(u, v)$ and $g(u, v)$ do not satisfy the properties assumed above. For instance, the null-clines $f = 0$ and $g = 0$ might have the form shown in Fig. 5.7.

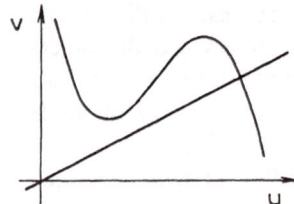

Fig. 5.7. Possible form of null-clines

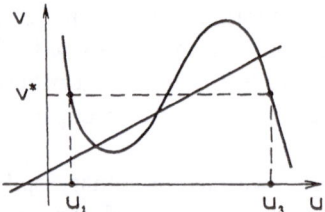

Fig. 5.8. Null-clines of a two-component activator-inhibitor system; v^* is the inhibitor concentration at the standing interface

In these cases we can try to reduce the problem to the already analyzed one by introducing new "activator" and "inhibitor" variables. In the case of Fig. 5.7 this can be done if we go to the new variables $u' = \text{const} - u$ and $v' = \text{const} - v$, thereby inverting the null-clines.

Finally, we can discuss the properties of spike patterns in two- and three-dimensional media. Then any individual spike becomes a small circular or spherical drop with high activator concentration which is surrounded by a larger region with increased concentration of the inhibitor. Such spikes form regular periodic lattices (with a spatial period not exceeding L_{min}).

If an active medium possesses a uniform stationary state, solitary spikes are possible. Under the same condition, one can produce spike lattices with an arbitrarily large spatial period. Interactions between the individual spikes become exponentially weak when the distance between them considerably exceeds the diffusional length L of the inhibitor. This implies that not only periodic spike lattices, but any irregular sparse spike distribution will also be practically stable.

5.3 Broad Strata

While the prototype of a spike is a critical nucleus in a one-component bistable medium, stable broad strata are most closely related to interfaces in such media.

As shown in Sect. 2.2, the propagation velocity of trigger waves depends on the parameters of a given medium. At a certain value of the control parameter (which is determined by the condition that the integral (2.2.6) vanishes) this velocity becomes equal to zero, and hence instead of the propagating trigger wave we have a standing interface.

In two-component activator-inhibitor systems, described by (5.2.1), the role of the control parameter in the activator equation is played by the local concentration v^* of the inhibitor at the interface. The interface is standing if the condition

$$\int_{u_1}^{u_3} f(u, v^*)\, du = 0 \tag{5.3.1}$$

is satisfied, where u_1 and u_3 are the roots of the equation $f(u, v^*) = 0$ that lie on the left and the right branches of the null-cline $f(u, v) = 0$ respectively (see Fig. 5.8).

This condition determines the value v^* of the inhibitor concentration at the standing interface.

The form of such a narrow interface can be found as a solution of the equation

$$f(u, v^*) + l^2 \partial^2 u / \partial x^2 = 0 \tag{5.3.2}$$

with the boundary conditions

$$u(x) \rightarrow u_1 \text{ for } x \rightarrow +\infty$$
$$u(x) \rightarrow u_3 \text{ for } x \rightarrow -\infty . \tag{5.3.3}$$

The width of this narrow front is thus of the same order of magnitude as the (small) diffusional length l of the activator.

If the inhibitor concentration were constant throughout the medium, the distribution of the activator would have been uniform outside the narrow interface region (approaching values u_1 and u_3 on the opposite sides of the interface). Actually, inhibitor concentration v varies outside the front, but with a characteristic diffusional length L which is much larger than the width of the interface.

In the regions outside the interface, the concentration u of the activator adjusts to the local inhibitor concentration v. Therefore, in these regions we have $u = u_{1,3}(v)$ where $u_1(v)$ and $u_3(v)$ are solutions to $f(u, v) = 0$ corresponding to the left and the right branches of the null-cline (Fig. 5.8).

The stationary inhibitor distribution is described by the equation

$$g\big(u_{1,3}(v), v\big) + L^2 \partial^2 v / \partial x^2 = 0 , \tag{5.3.4}$$

Suppose that the interface is located at $x = L_1$ and that the one-dimensional medium occupies the segment $(0, L_0)$. Then we can take the branch $u_1(x)$ on the right side (for $x > L_1$) and the branch $u_3(x)$ on the left side (for $x < L_1$), solving (5.3.4) separately in the intervals $(0, L_1)$ and (L_1, L_0).

The boundary conditions for (5.3.4) are

$$\partial v / \partial x = 0 \text{ for } x = 0 ,$$
$$v = v^* \text{ for } x = L_1 , \tag{5.3.5}$$
$$\partial v / \partial x = 0 \text{ for } x = L_0 .$$

The activator distribution $u(x)$ is obtained by matching the narrow interface at $x = L_1$ with slow variations $u = u_{1,3}\big(v(x)\big)$ in left and right outside regions.

The stationary dissipative pattern (Fig. 5.9) which consists of a standing interface and two outside regions with slow variations is called a *kink* structure. Obviously, a mirror reflection of any such structure will be again a valid stationary solution, so there are opposite kinds of kinks.

Combination of two opposite kinks (Fig. 5.10) produces a *stratum* of width $2L_1$ located in the center of the segment of length $2L_0$. This stationary solution can be infinitely extended to produce a periodic sequence of strata, with a sparial period $2L_0$.

When the length L_0 of the segment goes to infinity, we come to a solitary kink in an infinite medium. Clearly, such a solution is possible only in media which

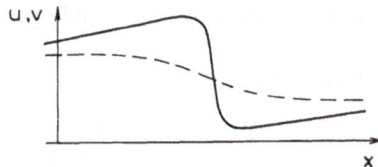

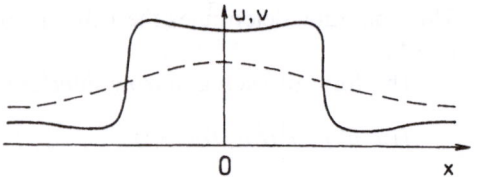

Fig. 5.9. Distributions of the acitvator (*solid line*) and the inhibitor (*dashed line*) in a kink pattern

Fig. 5.10. Distributions of the activator (*solid line*) and the inhibitor (*dashed line*) in a broad stratum

consist of bistable elements. On the other hand, if we expand the segment including a stratum we can produce a solitary stratum in an infinite medium. This solution is possible in active media that are made of excitable or bistable elements. If the medium consists of oscillatory elements it can support only periodic strata solutions.

When the distance between kinks is large enough their interaction is very weak. Therefore, in addition to regular periodic patterns one can produce also irregular structures that consist of strata of varying widths separated by varying intervals. Such irregular (or spatially stochastic) patterns will be practically stable (see [5.10]).

In two- and three-dimensional media the broad strata produce mosaic domain patterns. If an active medium permits the existence of a stable stationary uniform state, single domains are possible. A detailed analysis of the properties of broad strata and their domain patterns can be found in a review by *Kerner* and *Osipov* [5.8].

6. Neural Networks

Neural networks are a special class of distributed active systems that consist of discrete two-state elements linked by long-range activatory and inhibitory connections. The applicability of these formal models to description of the actual neural net of the brain remains doubtful. They give what is probably no more than a rough sketch of the extremely complex processes involved in the operation of the brain. Nevertheless, one can easily construct artificial active networks with these properties. Furthermore, artificial neural networks can be used to perform analog information processing. To use them in this way, one needs to be able to engineer the networks to give them the desired activity patterns.

Neural networks, which are designed to serve as devices with an associative memory, usually have a simple dissipative dynamics that results in relaxation to one of their possible stationary activity patterns. Here the principal problem is how to construct a network with the prescribed stationary patterns of activity. Other neural networks are used to store and retrieve the temporal sequences of patterns. In this case one needs to construct a network with many prescribed limit cycles of changing patterns.

Instead of purposefully engineering networks, we can use networks which are able to learn from experience. In the process of learning, such a network modifies the properties of connections between its neurons and thus gradually builds up its internal structure to perform the desired tasks.

6.1 Automata, Spins, and Neurons

The human brain is a giant network that consists of billions of neural cells, or *neurons*. Each neuron might have thousands of connections. In the simplest description, a neuron can be in any one of three different states, of rest, excitation, or refractoriness. Transitions between these states are controlled both by the processes inside the cell and by the electrical signals that arrive from other cells through the connections. A neuron becomes excited when its potential, which is obtained by summing the signals arriving from all its connections and averaging them over a suitable time interval, exceeds a certain threshold. After it has spent some time in the state of excitation, the neuron goes into the refractory state. This state has a very high excitation threshold, so that the neuron does not respond to arriving signals. After some time the threshold drops down and the neuron returns to the initial state of rest.

Connections between neurons are realized by *axons* and *dendrites*. An axon plays the role of the neuron output and carries an electrical signal to other cells. Dendrites represent the multiple inputs of a neuron. A transition into the excited state triggers an excitation pulse that propagates outwards along the axon at a speed of about 1 to 100 m/s. There are no direct electrical contacts between neurons. The transfer of a signal from the axon of one cell to the dendrite of another neuron takes place in a special location, the *synapse*, where the protrusions of both cells come close one to another; such transfer involves complicated chemical mechanisms. There are different kinds of synapses. Some of them are *stimulatory* (or *activatory*), and others are *inhibitory*. The latter produce in the dendrites the electrical signals of reverse polarity; when such signals arrive they *raise* the excitation threshold of a neuron. Since a single neuron has many connections, it can receive simultaneously a large number of signals of both polarities. Depending on the sign and the magnitude of the average potential induced by these signals, they can either trigger a transition into the excited state or keep the neuron in the state of rest.

The above brief (and greatly simplified) discussion shows that individual neurons can be modeled as excitable elements (cf. Chap. 3). However, neural networks are different from the homogeneous excitable media which were studied in Chap. 3. The principal difference is that any element of a neural network has a large number of long-range connections with other elements of the same network, while all interactions between individual elements in an excitable medium are local. Another difference is that the connections between elements are not identical: some of them are activatory, the others are inhibitory. Moreover, connections can differ also in their strengths. Hence, the neural network is strongly inhomogeneous, both because of a complicated topology and because of the varying properties of connections between individual elements.

As it turns out, these differences are of a fundamental character, leading to new kinds of behavior not found in excitable media. To investigate them one can, in principle, proceed from a model of a network which is composed of excitable elements. However, this model formulated in terms of three-state units remains too difficult for a fruitful analysis. Thus, further simplifications are usually made which reduce the neurons to *bistable* elements. This is achieved in the following way.

When the averaged electrical potential of a neuron, produced by many arriving pulses of different polarities, exceeds a certain threshold value, this neuron generates an excitation pulse which propagates outwards along its axon. If we keep the applied potential approximately constant, the neuron generates a periodic train of excitation pulses, separated by short intervals of refractoriness. When such train of pulses is being generated, we say that a neuron is in the *active* state. On the other hand, if the applied potential is below the threshold, the neuron remains in the *passive* state of rest.

Hence, we can specify the state of a neuron j by a binary variable s_j which takes the value 1 if this neuron is active within a given small interval of time and the value 0 if it is passive.

Suppose we have two neurons connected by an inhibitory synapse. Then, if the first neuron i is active ($s_i = 1$), it will induce an inhibiting potential in the second neuron j and the latter will stay passive ($s_j = 0$). The situation is changed when these neurons are connected by an activatory synapse. Then activity of one neuron stimulates the activity of the other, and both of them stay in the active state $(s_i = s_j = 1)$.

It is convenient to characterize a synaptic connection by its *weight* G_{ij} which is taken as positive for an activatory synapse and negative for an inhibitory one; the absolute value of G_{ij} specifies the strength of a connection. If $G_{ij} = 0$, a connection between this pair of neurons is absent. The set of weights G_{ij} for all pairs of neurons constitutes the *synaptic matrix* $[G_{ij}]$ of a given network.

Since pulses propagate along the connections at a finite speed, there is a certain *delay* between the moment when a pulse is generated and the moment when it reaches another neuron. However, in the simplest model we can neglect such delays. Then the potential which is induced on a given neuron by all other neurons in the network can be written (employing some arbitrary units) as

$$U_i = \sum_j G_{ij}\, s_j \, , \tag{6.1.1}$$

If the threshold (or the *bias*) of neuron i is B_i, it will become active when $U_i \geq B_i$ and passive when $U_i < B_i$. Hence, the new state s'_i of this neuron, produced by application of potential (6.1.1), will be

$$s'_i = H\left(\sum_k G_{ij}\, s_j - B_i\right) \tag{6.1.2}$$

where $H(z)$ is the step function, $H(z) = 1$ for $z \geq 0$ and $H(z) = 0$ for $z < 0$.

The above model of a neural network, with formal neurons represented as binary units, was first proposed by *McCulloch* and *Pitts* [6.1] in 1943. This model defines the dynamics of a network as a series of transitions (6.1.2) determined by the momentary states of activity of all its neurons and the weights of all its synaptic connections.

To make the definition complete, one should specify also the *sequence* of transitions, i.e. the order in which the states of the neurons are updated. In realistic neural networks this order is determined by delays in the arrival of signals from other neurons. Since we neglect such delays, some convention should be adopted.

In effect, different updating schemes can be employed. For instance, in a sequential scheme we pick one neuron at random and change its state according to rule (6.1.2), then pick another one, calculate the new value of the potential and use (6.1.2) again, etc. Another possibility is synchronous parallel updating. In this scheme, by using the states of neurons at a given time step n, we find the new states of all neurons at the next discrete step $n + 1$, i.e. we put

$$s_i^{n+1} = H\left(\sum_k G_{ij} s_i^n - B_i\right) \tag{6.1.3}$$

for all neurons i.

Despite its simple definition, the McCulloch-Pitts model of a neural network has extremely rich behavior. Depending on a particular synaptic matrix, such a network can possess very complicated steady activity patterns, display complex periodic oscillations, etc. These features make it very promising for the purposes of analog information processing.

Although the McCulloch-Pitts model with formal neurons was originally proposed to describe the neural networks of the human brain, it is still very difficult to answer whether this model is even a valid first approximation for brain studies. Realistic neurons are much more complicated than the two-state units employed in this model. At present, there are some indications that individual neurons can perform complex operations of information processing, i.e. that any single neural cell may represent an entire biological computer (see [6.2, 3]. Moreover, the brain has many specialized neurons with different properties. Finally, an important role in the brain operation is probably played by humoral chemical substances that propagate in the brain tissue.

However, all these doubts do not cool an enthusiasm for studies of formal neural networks. The reason for this lies in the fact that we can easily construct *artificial* networks with the required properties. Indeed, what we need is only a set of bistable active elements linked by adjustable connections. Every such element represents a simple *automaton* which changes its state by a simple updating rule (6.1.2), in accordance with the signals coming from all other elements. This updating rule constitutes the transition algorithm for such set of automata.

Biological synapses can pass signals in only one direction, i.e. from an axon to a dendrite. Therefore the elements G_{ij} and G_{ji} of the synaptic matrix correspond in this case to two different connections: from the neuron j to the neuron i and in the reverse direction. Hence, there are no general reasons to expect that these two quantities should coincide in a realistic neural network. Generally, the synaptic matrix $[G_{ij}]$ is asymmetric. However, in artificial neural networks we can easily implement symmetric connections. It turns out that the networks with symmetric synaptic matrixes are more amenable to a theoretical analysis.

If $G_{ij} = G_{ji}$ for every pair (i, j) of neurons, we can introduce a function

$$E = -\frac{1}{2} \sum_{i,j} G_{ij} s_i s_j + \sum_i B_i s_i , \qquad (6.1.4)$$

that plays a role of the effective "energy" of such neural network. To explain this interpretation of E, let us calculate the difference of its values before and after a single act of updating (6.1.2):

$$\Delta E = E' - E = -\sum_j G_{ij}(s_i' - s_i)s_j + B_i(s_i' - s_i)$$

$$= -\left(\sum_j G_{ij} s_j - B_i\right)(s_i' - s_i) . \qquad (6.1.5)$$

It can be shown that ΔE is nonpositive. Indeed, if $s_i = 1$ and $s_i' = 0$, it follows from (6.1.2) that $\sum_j G_{ij} s_j < B_i$ and $\Delta E < 0$. If $s_i = 0$ and $s_i' = 1$, we have $\sum_j G_{ij} s_j > B_i$ and therefore $\Delta E < 0$. Moreover, $\Delta E = 0$ only if a transition did not occur, i.e. if $s_i' = s_i$. Hence, the evolution of activity in such neural network is accompanied by a steady decrease of E, i.e. the network has simple dissipative dynamics.

The energy (6.1.4) cannot decrease indefinitely because it is bounded from below:

$$E \geq -\sum_{i,j} |G_{ij}| . \qquad (6.1.6)$$

When a minimum of E is reached, further transitions become impossible. This implies that evolution of such a network always leads to some steady activity pattern that corresponds to a minimum of energy E.

The physical analogy can be continued. It is possible to specify a system of interacting spins with the same dynamics.

A spin is an internal magnetic moment of a particle, which is a vector with a certain direction. Below we consider idealized Ising spins that can point only "up" and "down" a certain axis. Such spins are described by binary variables $S_i = \pm 1$ (+1, if a spin is aligned upwards, and −1 in the opposite case). The energy E of these interacting spins is given by an expression

$$E = -\frac{1}{2} \sum_{ij} J_{ij} S_i S_j + \sum_i b_i S_i , \qquad (6.1.7)$$

where J_{ij} are the coupling coefficients and b_i is the external magnetic field acting on spin i. At zero temperature, when thermal fluctuations are absent, this system relaxes to a state with the minimal energy. Relaxation occurs by a sequence of spin-flips $S_i \rightarrow S_i'$ such that

$$S_i' = \text{sign} \left(\sum_j J_{ij} S_j - b_i \right) . \qquad (6.1.8)$$

Each spin-flip results in a decrease of the total energy:

$$\Delta E = E' - E = -\sum_j J_{ij}(S_i' - S_i)S_j + b_i(S_i' - S_i)$$

$$= -2 \left(\sum_j J_{ij} S_j - b_i \right) \text{sign} \left(\sum_j J_{ij} S_j - b_i \right) \leq 0 . \qquad (6.1.9)$$

We have taken into account that, if a spin-flip occurred, $S_i = -S_i'$.

To establish a correspondence between a neural network and a system of Ising spins, we map active states of neurons into the "up" states of spins and passive states into the "down" states. Putting $s_i = (1/2)(S_i + 1)$ in (6.1.4) yields

$$E = -\frac{1}{8} \sum_{i,j} G_{ij} S_i S_j - \frac{1}{4} \sum_{i,j} G_{ij} S_j - \frac{1}{8} \sum_{i,j} G_{ij}$$

$$+ \frac{1}{2} \sum_i B_i S_i + \frac{1}{2} \sum_i B_i \qquad (6.1.10)$$

$$= E_0 - \frac{1}{8} \sum_{i,j} G_{ij} S_i S_j + \frac{1}{2} \sum_i \left(B_i - \frac{1}{2} \sum_i G_{ij} \right) S_i ,$$

where

$$E_0 = -\frac{1}{8} \sum_{i,j} G_{ij} + \frac{1}{2} \sum_i B_i . \qquad (6.1.11)$$

Hence, we see that, when the "energy" of a neural network is expressed in terms of spin variables S_i, it reduces to the energy (6.1.7) of the spin system. To complete the reduction, we should take $(1/4)G_{ij}$ as the coupling coefficient J_{ij} in (6.1.7) and should assume that the external magnetic fields in (6.1.7) are

$$b_i = \frac{1}{2} \left(B_i - \frac{1}{2} \sum_j G_{ij} \right) . \tag{6.1.12}$$

The presence of a constant correction E_0 in (6.1.10) does not influence the dynamics.

Because of this one-to-one correspondence between a network of formal neurons and a system of Ising spins, in theoretical studies these two descriptions are freely alternated. Often it is simply assumed that the states of neurons are specified by spin-like variables S_i, so that $S_i = +1$ in the active state and $S_i = -1$ in the passive state of neuron i. Then the updating algorithm coincides with (6.1.8), the formal "energy" of this network is given by (6.1.7), and the coupling coefficients J_{ij} are called the weights of connections. Below we use this interpretation unless otherwise specified. From a mathematical point of view, McCulloch-Pitts neurons and Ising spins represent the same discrete binary automata.

In the next section we show how the formal neural network with symmetric connections can be used to realize the associative recall of stored patterns, which is a fundamental problem of analog information processing.

6.2 Associative Memories

We say that a system possesses the property of *associative memory* if it keeps in its memory several prototype patterns and, when a distorted version (or a fragment) of one of these patterns is presented, it is able to retrieve the corresponding prototype automatically. Hence, such a system performs classification of input patterns, assigning each of them to one of the stored prototypes. In this sense, associative memory also represents a special case of *pattern recognition.*

One of the possible ways to realize an analog associative memory is to construct a distributed dynamical system or a discrete network whose attractors in the configuration space coincide with the prototype patterns. Then every such pattern has its own basin of attraction. Any initial condition which represents one of the permitted patterns falls into a certain attraction basin. In the course of time, such an initial pattern is then transformed into the attractive pattern of this basin, i.e. into one of the memorized prototypes.

Therefore, by applying a certain pattern as an initial condition to such a distributed active system, we can perform its automatic (i.e. analog) classification and retrieval of the nearest prototype. Thus we come to the problem of the purposeful design (or *engineering*) of distributed dynamical systems with given attractors.

In this section we consider a particular realization of a dynamical system with associative memory proposed by *Hopfield* [6.4]. Since this work was stimulated by theoretical studies of special physical systems called *spin glasses*, we begin with a brief discussion of their relevant features.

Suppose we have a system of interacting Ising spins with an energy

$$E = -\frac{1}{2} \sum_{i,j} J_{ij} S_i S_j , \qquad (6.2.1)$$

and the dissipative dynamics (6.1.8). Let us assume first that all coefficients J_{ij} are positive. Then in the state with the minimal energy E all spins will have the same orientation, either $S_i = +1$ or $S_i = -1$ for all i. Such a state of a system of interacting spins is called "ferromagnetic". Obviously, the interaction energy $\delta E_{ij} = -J_{ij} S_i S_j$ of any spin pair (i, j) then achieves its minimal possible value $\delta E_{ij} = -J_{ij}$.

In a spin glass the coefficients J_{ij} are random and take both positive and negative values. In this case in a state with the minimal energy E it is usually impossible to minimize interaction energies of *all* spin pairs for the same spin configuration.

There will be many spin triplets for which the product $J_{ij} J_{jk} J_{ki}$ is negative. Let us try to minimize for such a triplet all three interaction energies $\delta E_{ij} = -J_{ij} S_i S_j$, $\delta E_{jk} = -J_{jk} S_j S_k$, and $\delta E_{ki} = -J_{ki} S_k S_i$. If we take $S_i = +1$, then, to make δE_{ij} minimal, we should choose $S_j = \text{sign}(J_{ij})$. Furthermore, the minimal value of δE_{jk} is reached if we choose $S_k = \text{sign}(J_{jk}) \text{sign}(J_{ij})$. But then we find $\delta E_{ki} = -J_{ki} \text{sign}(J_{jk}) \text{sign}(J_{ij}) > 0$ which is not the minimal possible value of the interaction energy of spins i and k.

Hence, the minimization conditions for interaction energies of different spin pairs are incompatible. A spin glass is an example of so-called *frustrated systems* with conflicting factors.

Because of frustration, a spin glass has many states with local minima of energy E that correspond to different stable spin configurations. We can say that, for a given random choice of coefficients J_{ij}, a spin glass keeps in its memory a large number of different stable patterns.

If we begin with an arbitrary initial pattern, it will relax to the stable spin pattern which is closest to it. Hence, a spin glass not only stores certain fixed patterns, but it is also able to classify other patterns according to their similarity with the stored ones.

Obviously, the same behavior will be found in neural networks with random symmetric synaptic matrices. Such networks have many steady states with different activity distributions that correspond to local energy minima.

The set of stored patterns in a spin glass is random. To obtain a *controllable* associative memory, we should find some prescription that allows us to construct the matrix $[J_{ij}]$ in such a way that our system will have a given set of prototype patterns as its minimal energy configurations.

Suppose that we want to memorize a pattern which is characterized by a particular value $S_i = \xi_i$ for every spin i. Let us choose the weights so that $J_{ij} = \xi_i \xi_j$. Then it can easily be seen that the stored spin pattern will correspond to a minimum of energy, i.e. that it will be an attractive pattern. Indeed, for any spin pair (i, j) the interaction energy is then $\delta E_{ij} = -J_{ij} \xi_i \xi_j = -\xi_i^2 \xi_j^2 = -1$, i.e. it achieves its minimal possible value.

Such a system is not yet able to classify patterns, since it keeps in its memory only one pattern to which any initial pattern relaxes[1]. However, this deficiency can be removed.

Suppose that we want to store M different patterns $\{\xi_j^{(m)}\}$, $m = 1, 2, \ldots, M$. For simplicity, let us assume for a moment that all these patterns are *orthogonal*, i.e. that the conditions

$$\frac{1}{N} \sum_{j=1}^{N} \xi_j^{(m)} \xi_j^{(m')} = \delta_{mm'} \qquad (6.2.2)$$

are satisfied.

Following *Hebb* [6.5], we can choose weights J_{ij} thus:

$$J_{ij} = \frac{1}{N} \sum_{m=1}^{M} \xi_i^{(m)} \xi_j^{(m)} . \qquad (6.2.3)$$

Then all M patterns will be among the steady states of this system (i.e. of an artificially constructed "spin glass").

To prove this, we introduce internal fields h_i such that

$$h_i = \sum_{j=1}^{N} J_{ij} S_j \qquad (6.2.4)$$

and note that (in the absence of external fields) the transition algorithm (6.1.8) is simply

$$S_i' = \text{sign } h_i , \qquad (6.2.5)$$

Hence, for a stable stationary pattern, the directions of spins should coincide with the directions of internal fields, i.e. the condition $h_i S_i > 0$ should be satisfied.

If weights J_{ij} are given by (6.2.3), then in the state which corresponds to a stored pattern $\{\xi_i^{(m')}\}$ we have

$$h_i \xi_i^{(m')} = \frac{1}{N} \sum_j \sum_m \xi_i^{(m)} \xi_j^{(m)} \xi_i^{(m')} \xi_j^{(m')}$$

$$= \xi_i^{(m')} \xi_i^{(m')} = 1 > 0 . \qquad (6.2.6)$$

Hence, every stored pattern is indeed a stable stationary state of our system.

If all prototype patterns $\{\xi_i^{(m)}\}$ are orthogonal, the maximal number of stored patterns in the model can be easily estimated. Every such pattern can be considered as a certain vector in an N-dimensional euclidean space. Since the total number of orthogonal vectors in this space cannot be larger then its dimensionality, we immediately see that the maximal number of stored orthogonal patterns in a network of N formal neurons is N.

---

[1] More correctly, there are two attractive patterns, because the total energy is minimal for a mirror reflected pattern $S_i = -\xi_i$ as well.

The requirement (6.2.2) of rigorous orthogonality of recorded patterns imposes excessively severe restrictions on possible prototypes, which are difficult to satisfy in concrete applications. Fortunately, it turns out that, for large numbers N, any two randomly chosen patterns would usually be *almost* orthogonal and this approximate orthogonality is sufficient for operation of an associative memory based on the Hebb rule.

Suppose we have two random patterns represented by formal vectors $\{\xi_i^{(m)}\}$ and $\{\xi_i^{(m')}\}$, every component of which is an independent binary random variable that takes values +1 and −1 with equal probability. For such a pair of patterns, every quantity $r_i = \xi_i^{(m)} \xi_i^{(m')}$ is a random variable with two equiprobable values +1 and −1. Hence, the scalar product (or the *overlap integral*) of these two vectors

$$R_{mm'} = \sum_{i=1}^{N} \xi_i^{(m)} \xi_i^{(m')} \tag{6.2.7}$$

which enters into the orthogonality condition (6.2.2) is a sum of N independent random numbers r_i. Therefore at large N the variance of the random quantity $R_{mm'}$ is

$$\langle R_{mm'}^2 \rangle = N \ . \tag{6.2.8}$$

Consequently, when $N \to \infty$ an estimate

$$\frac{1}{N} \sqrt{\langle R_{mm'}^2 \rangle} = N^{-1/2} \tag{6.2.9}$$

holds. We see that in the limit of a large number of neurons any two randomly chosen patterns are usually almost orthogonal.

Small deviations from orthogonality lead to additional random variations in the fields h_i. To determine the intensity of such variations we can estimate h_i in a state which coincides with one of the stored patterns, i.e. $S_j = \xi_j^{(m')}$:

$$h_i = \sum_j J_{ij} \xi_j^{(m')}$$

$$= \xi_i^{(m')} + \frac{1}{N} \sum_{m \neq m'} \xi_j^{(m)} \xi_j^{(m')} \ . \tag{6.2.10}$$

The second term in (6.2.10) represents a contribution from all other stored patterns that arises because of their non-orthogonality. It plays the role of random noise imposed on the regular component of fields h_i given by the first term. To ensure the correct operation of an associative memory, this noise should be small, on average, in comparison to the regular field component.

To estimate this contribution, we multiply both sides in (6.2.10) by $\xi_i^{(m')}$ and sum over i. After averaging, this yields

$$\frac{1}{N} \left\langle \sum_i h_i \xi_i^{(m')} \right\rangle = 1 + (1/N^2) \sum_{m \neq m'} \langle R_{mm'}^2 \rangle \ . \tag{6.2.11}$$

Hence, the magnitude of the contribution from the noise component of the field is about M/N and we can expect that the reliable operation of such a network as an associative memory device is maintained until $M \ll N$. A more sophisticated analysis undertaken by *Amit* et al. [6.6] for uncorrelated random patterns shows that the breakdown of operation is achieved at $M/N = 0.14$, when the retrieval quality is sharply deteriorated. For smaller numbers M of stored patterns, the system can occasionally retrieve a wrong (spurious) pattern, but the probability of this remains insignifcant.

The ratio of the maximal number of stored and correctly retrieved patterns to the total number of neurons is the *memory capacity* g of the network. We see that in the case of uncorrelated random patterns the memory capacity is $g = 0.14$, if the Hebb rule for construction of the synaptic matrix is used. This is smaller almost by a factor of ten than the memory capacity $g = 1$ which is achieved under the same rule in the case of orthogonal patterns.

These estimates are very sensitive to the assumption of complete randomness of all stored patterns or, more precisely, to the assumption that every element of such a pattern is an independent random variable with a zero mean value. If the stored patterns are correlated, the intensity of the noise contribution in fields h_i becomes larger and the probability of retrieval failure increases. This imposes a serious limitation because the patterns to be stored often belong to a certain class, and hence are not completely random. Similar difficulties arise when the stored patterns are *biased*, i.e. when $\left\langle \xi_i^{(m)} \right\rangle \neq 0$. To improve the performance of the network in such situations, we can try to replace the Hebb rule (6.2.3) by something better.

One of the possibilities is a *pseudo-inverse rule* discussed by *Kohonen* [6.7] and *Personnaz* et al. [6.8]. To apply this prescription, the matrix of overlap integrals $R_{mm'}$ between all prototype patterns should first be calculated and the inverse matrix $(R^{-1})_{mm'}$ determined. Then the synaptic matrix can be constructed using

$$J_{ij} = \sum_{m,m'} (R^{-1})_{mm'} \xi_i^{(m)} \xi_j^{(m')} . \tag{6.2.12}$$

Let us verify that a network with such a synaptic matrix actually stores all required patterns. Suppose that the network is in a state that corresponds to one of these patterns, $\{\xi_i^{(k)}\}$, and find the fields h_i in such a state:

$$
\begin{aligned}
h_i &= \sum_j J_{ij} \xi_j^{(k)} \\
&= \sum_{m,m'} (R^{-1})_{mm'} \xi_i^{(m)} \sum_j \xi_j^{(m')} \xi_j^{(k)} \\
&= \sum_m \xi_i^{(m)} \sum_{m'} (R^{-1})_{mm'} R_{m'k} = \sum_m \xi_i^{(m)} \delta_{mk} = \xi_i^{(k)}
\end{aligned}
\tag{6.2.13}
$$

Hence, $h_i \xi_i^{(k)} > 0$ and any recorded pattern $\{\xi_j^{(k)}\}$ represents a stable stationary state of the network.

Note that the synaptic matrix $[J_{ij}]$ given by (6.2.12) is a *projection matrix*. Each of the prototypes $\{\xi_j^{(k)}\}$ is its exact eigenvector, with eigenvalue 1; the other $N - M$

eigenvalues are all zero. In particular, this implies that the maximal number of stored patterns in this case is N, i.e. it coincides with the total number of spins (formal neurons) in the system. In other words, the memory capacity for the pseudo-inverse rule is $g = 1$.

The same matrix (6.2.12) can be constructed by an iterative learning rule (see [6.7,9]):

$$J_{ij}(t+1) = J_{ij}(t) + \frac{\varepsilon}{N} \sum_{m=1}^{N} \left(1 - h_i^{(m)} \xi_i^{(m)}\right) \xi_i^{(m)} \xi_j^{(m)} . \tag{6.2.14}$$

This rule means that at every discrete time step t we have to calculate fields $h_i^{(m)}$ for all M prototype patterns $\xi_i^{(m)}$ with a given matrix $J_{ij}(t)$ and then to modify the matrix elements in accordance with (6.2.14). The learning procedure starts with $J_{ij}(0) = 0$ and is continued until the condition $h_i^{(m)} \xi_i^{(m)} = 1$ is satisfied for all M patterns.

It is possible to estimate (see the review by *Kinzel* [6.10]) how quickly iterations (6.2.14) converge to their fixed point, i.e. to matrix (6.2.12). The learning time (i.e. the number of required iterations) is minimal for an optimal value of the parameter ε, given by

$$\varepsilon_{\text{opt}} = (1 + M/N)^{-1} . \tag{6.2.15}$$

Then it is equal to

$$t_{\min} = \left[\ln \frac{1 + M/N}{2(M/N)^{1/2}} \right]^{-1} . \tag{6.2.16}$$

These estimates are derived for random, uncorrelated and unbiased patterns in the limit when $N \to \infty$ and $M \to \infty$ but the ratio M/N is fixed. We see that the learning time diverges when the total number M of prototype patterns comes closer to N, i.e. if we try to make use of the full memory capacity $g = 1$ of such network.

The memory capacity yielded by rules (6.2.12) or (6.2.14) is still not the best possible. *Gardner* [6.11] investigated the question of what is the maximal memory capacity that can be achieved by the best choice of the learning rule, both for biased and unbiased random patterns. She found that the best memory capacity for unbiased patterns is $g_{\max} = 2$. The maximal memory capacity for biased patterns was also estimated in [6.11, 12].

The best possible memory capacity can be reached by using the *perceptron rule* (see [6.7, 9, 13–15]):

$$J_{ij}(t+1) = J_{ij}(t) + \frac{\varepsilon}{N} \sum_{m=1}^{M} \xi_i^{(m)} \xi_j^{(m)} H \left(1 - h_i^{(m)} \xi_i^{(m)}\right) \tag{6.2.17}$$

where $H(z)$ is a step function, and ε is a small parameter that controls the rate of learning.

The iterative algorithm (6.2.17) is proved by *Krauth* and *Meyard* [6.14] to converge within a finite number of steps if there is a solution, i.e. if the network is able to memorize a given set of prototype patterns. For random unbiased patterns (see [6.16]) the learning time diverges as $(2 - M/N)^{-1}$ at the maximal storage capacity $g = 2$.

The weight coefficients J_{ij} that are obtained by using above mentioned rules can take a continuous range of positive or negative values. This is not very convenient from the point of view of potential applications. Therefore, the original weight coefficients are often clipped, i.e. the genuine synaptic matrix $[J_{ij}]$ is replaced by another matrix with the elements

$$J'_{ij} = \text{sign}(J_{ij}) \tag{6.2.18}$$

which can take only values ± 1. Clipping does not introduce drastic distortions into the operation of associative memories. In the case of the Hebb learning rule (6.2.3) and random unbiased patterns it decreases the memory capacity of a network from 0.14 to 0.11 (see [6.17, 18]).

Remarkably, the property of associative memory is maintained, as shown by *Derrida* et al. [6.19] and by *Tsodyks* and *Feigelman* [6.20], even if we delete a considerable fraction of connections, i.e. if we put $J_{ij} = 0$ for some randomly chosen pairs (i, j) of elements[2]. When some connections are randomly deleted, this leads to an effective noise contribution in the fields h_i. However, this contribution would be given by a sum of many independent random terms that almost cancel one another.

The Hopfield model which was discussed above can also be formulated in terms of a dynamical network with continuous variables.

Let us consider, for instance, a dynamical system that is described by the following set of equations for continuous variables σ_i:

$$\dot{\sigma}_i = -\frac{\partial \mathcal{E}}{\partial \sigma_i}, \quad i = 1, 2, \ldots, \quad N, \tag{6.2.19}$$

where function $\mathcal{E}$ is defined as

$$\mathcal{E} = -\sum_{i,j} J_{ij} \sigma_i \sigma_j + \lambda \sum_i (\sigma_i^2 - 1)^2. \tag{6.2.20}$$

Any stable attractive state of this dynamical system corresponds to a minimum of $\mathcal{E}$. In the limit $\lambda \to \infty$, minima of $\mathcal{E}$ can be located only at the points where the values of σ_i are either $+1$ or -1. But at such points function $\mathcal{E}$ coincides with the energy (6.2.1) for a system of N spins. Therefore, all results that are derived for networks with binary variables are also valid for asymptotic states of such dynamical systems.

Points where all variables σ_i have values ± 1 are the vertices of an N-dimensional hypercube. Some of these vertices correspond to stable stationary states of our dynamical system. Every such fixed point is an attractor which has its own basin of

---

[2] If we break the respective connections leading from an element i to an element j and in the reverse direction independently, this results in the asymmetric synaptic matrix $(J_{ij} \neq J_{ji})$. In this case no energy function E can be constructed. Nevertheless, the network dynamics continues to be determined by (6.2.5).

attraction. Since the dynamical system (6.2.19) has no other attractors than these fixed points, (almost) the whole configuration space is divided into their basins.

If every attractive fixed point in the configuration space corresponds to one of the stored patterns, this dynamical system is able to perform analog pattern recognition and retrieval of the nearest stored pattern, i.e. it has the property of associative memory. To perform recognition, we should apply the presented pattern as an initial condition and follow the subsequent evolution of our dynamical system.

The above example illustrates a principal possibility of analog associative memory based on a dynamical system with continuous variables, but it is not very convenient for practical implementation. Below we describe another example, proposed by *Hopfield* and *Tank* [6.21], that can be easily realized as an electronic circuit.

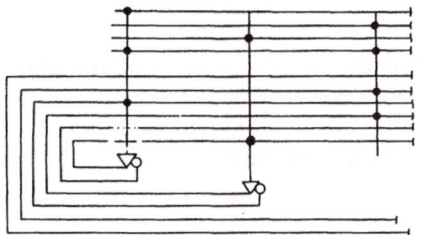

Fig. 6.1. Scheme of a circuit that implements the Hopfield model. Triangles are nonlinear amplifiers; their inverted outputs are shown as white circles. Black dots are the resistances realizing connections between the inputs and the outputs of different amplifiers

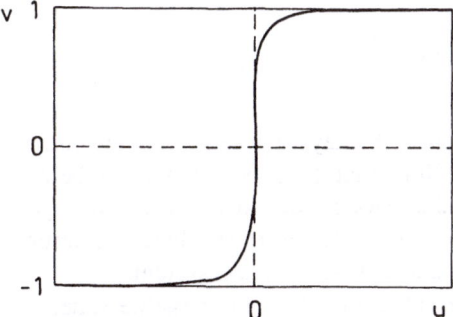

Fig. 6.2. Step-like transfer function

Let us consider a network of connected nonlinear amplifiers (Fig. 6.1). For any amplifier the input (u_i) and the output (v_i) voltages are related by a transfer function $v = g(u)$ that is monotonic and bounded (Fig. 6.2). We choose the reference value and the units of measurement for the output voltage v in such a way that it varies within the bounds $+1$ and -1. The internal resistance of an amplifier i is ϱ_i and its electrical capacity is C_i. The input of any amplifier consists of electrical signals coming from all other amplifiers through the connection resistances ϱ_{ij}. In addition to a normal output, every amplifier possesses an inverted output that produces the same signal but with inverted polarity. When the connection between elements i and j is characterized by a positive weight coefficient J_{ij}, the input of the amplifier i should be taken from the normal output of the amplifier j, so that $J_{ij} = 1/\varrho_{ij}$. A connection with the negative weight coefficient $J_{ij} = -1/\varrho_{ij}$ can be realized by taking the input signal from the inverted output of jth amplifier. The dynamics of input voltages in such a system is described by the equations

$$C_i \dot{u}_i = \sum_j J_{ij} v_j - u_i/R_i \, ,$$

$$v_j = g(u_j) \, , \tag{6.2.21}$$

where the effective resistance R_i is

$$1/R_i = 1/\varrho_i + \sum_j 1/\varrho_{ij} \, . \tag{6.2.22}$$

Let us consider function $\mathcal{E}$ defined as

$$\mathcal{E} = -(1/2) \sum_{i,j} J_{ij} v_i v_j + \sum_j (1/R_j) \int_0^{v_j} g^{-1}(v) \, dv \, , \tag{6.2.23}$$

where $u = g^{-1}(v)$ is the inverse of $v = g(u)$. Its time derivative is

$$d\mathcal{E}/dt = -\sum_{i,j} J_{ij} v_i \dot{v}_j + \sum_j (1/R_j) \dot{v}_j g^{-1}(v_j)$$

$$= -\sum_j \left(\sum_i J_{ij} v_i - u_j/R_j \right) \dot{v}_j \tag{6.2.24}$$

$$= -\sum_j C_j \dot{u}_j \dot{v}_j = -\sum_j C_j (dg/du_j)(\dot{u}_j)^2 \, .$$

Since dg/du is positive, $d\mathcal{E}/dt \leq 0$. Hence, $\mathcal{E}$ plays a role of the Lyapunov function for differential equations (6.2.21). Note that $\mathcal{E}$ is bounded from below because the second term in the expression (6.2.23) for $\mathcal{E}$ goes to $+\infty$ when voltages v_j reach their minimal (-1) or maximal ($+1$) values. Therefore any local minimum of this function corresponds to some stable stationary state of the system.

Since every variable v_j varies in the interval from -1 to $+1$, possible states of this system lie in an N-dimensional hypercube $\{-1 < v_j < +1\}$. If the transfer function $v = g(u)$ is nearly step-like, the last term in (6.2.23) differs from zero only in the vicinity of the hypercube borders. Inside the hypercube, the dominant contribution to $\mathcal{E}$ is given by the first term in (6.2.23). But this quadratic term can reach a minimum in the interior of the hypercube only if the matrix $[J_{ij}]$ is negatively defined. In our case the opposite is true and hence all minima should lie on the borders of the hypercube. If we require additionally that $J_{ij} = 0$, all minima will lie only at the vertices of this hypercube, i.e. they will be given by the sets $\{v_j\}$ that consist only of $+1$ and -1. Hence, when function $g(u)$ approaches the step form, all stable stationary states of this dynamical system will be the same as in the system of spins with the interaction energy (6.2.1). If we construct the matrix $[J_{ij}]$ by using the Hebb rule or other learning algorithms, this dynamical system (i.e. an electronic circuit) will be endowed with associative memory. When a new pattern is input into the system as its initial condition, the nearest prototype pattern is automatically identified and retrieved after a certain relaxation time.

Table 6.1.

John Stewart Denker	8128
Lawrence David Jacke	17773
Richard Edwin Howard	5952
Wayne P. Hubbard	7077
Brian W. Straughn	3126
John Henry Scofield	8109

Denker [6.22] used this system of analog associative memory to memorize the names and the telephone numbers of six people from his laboratory (Table 6.1). Each entry was 25 characters long, and was encoded into a binary form by using 5 bits per character. Therefore each entry was represented by a vector $\{\xi_i\}$ which was 125 bits long. The synaptic matrix $[J_{ij}]$ was constructed by using the simplest Hebb rule. The dynamics of voltages in the electronic circuit was simulated by a computer that numerically integrated equations (3.3.22).

Table 6.2.

Time	Energy		
0	0	john s	
0.20	−0.0784	john sdewirubneoimv	8109
0.40	−0.8426	john sdewirtbnenimw	8129
0.60	−0.8451	john sdewirtbnenimv	8129
0.80	−0.8581	john sdewirt nenkmv	8128
1.00	−0.9099	john sdewart denker	8128
1.20	−0.9824	john stewart denker	8128

Table 6.3.

Time	Energy		
0	0	john h	
0.20	−0.0665	john hdnwybqbofmalt	8109
0.40	−0.8579	john henwybqcofield	8109
0.60	−0.9094	john henry scofield	8109

Table 6.2 shows the subsequent temporal evolution of an initial vector that corresponds to the entry "john s" (time and energy are measured in arbitrary units). We see that even this small piece is sufficient to reconstruct correctly the entire entry. When another initial condition "john h" is specified, which differs from the previous case in only one symbol, the result of the evolution is categorically changed (Table 6.3); now this system correctly retrieves the name and the telephone number of another person.

The abilities of the system are not restricted to recovery of the omitted symbols. It can also identify an entry from its *distorted* fragment. Table 6.4 shows the memory operation when the input word was "larry".

Table 6.4.

Time	Energy	
0	0	larry
0.20	−0.0213	larrynce david jacke17773
0.40	−0.7977	larrynce david jacke17773
0.60	−0.8713	larrynce david jacke17773
0.80	−0.8813	lasrence david jacke17773
1.00	−1.001	lasrence david jacke17773
1.20	−1.001	lawrence david jacke17773

Table 6.5.

Time	Energy	
0	0	garbage
0.20	−0.0244	garbagee lafj naabd 5173
0.40	−0.6280	garbaged derjd naabd 7173
0.60	−0.6904	garbaged derjd naabd 7173
0.80	−0.6904	gasbafed derjd naabd 7173
1.00	−0.7595	gasbabed derjd naabd 7173
1.20	−0.7709	fasjebad derjd naabd 7173
1.40	−0.8267	fasjebad derjd naabd 7173
1.60	−0.8282	fasjeb d derjd naabd 7173

The six entries used in the experiment [6.22] were not orthogonal. Therefore, besides the stored prototypes, the system had also some spurious attractive patterns. Evolution resulting in one of these spurious patterns is displayed in Table 6.5.

6.3 Complex Combinatorial Optimization

In many applications one is faced with the task of choosing the optimal variant among a giant number of possibilities. The typical example of such task is the so called *travelling salesman problem*.

Suppose that we have N points ("cities") A, B, C, D, ... , that are scattered at random on a plane and we know all the distances $l_{AB}, l_{AC}, \ldots , {}_{BC}, \ldots$, between them. The problem is to find the shortest closed tour that connects all these points.

Let us estimate the total number of variants in this problem. By choosing a sequence of points B, F, E, G, D, ... , W we obtain a tour with the total length

$$l = l_{BF} + l_{FE} + l_{EG} + l_{GD} + \ldots + l_{WB} . \tag{6.3.1}$$

There are $N!$ sequences but some of them define the same tours. Indeed, the initial point and the direction of the round are arbitrary, and hence every tour corresponds to $2N$ different sequences. Hence, the number of variants is $N!/2N$.

If we try to solve the travelling salesman problem with a computer by an exhaustive search, the total number of required computer operations will not be smaller than the number of available variants. But this number grows as a factorial, i.e. faster than any power of N. When the number of cities increases, the amount of computer time required to find an optimal route diverges so rapidly that an exact solution of this problem by an exhaustive search is impossible. Moreover, it is proved that there is no shorter algorithm for an exact solution with a number of operations given by some power of N.

The above example is only a special case of a class of problems with *nonpolynomial complexity*, or *NP-problems* (see [6.23]). For such combinatorial optimization problems there are no algorithms leading to an exact solution in a number of steps proportional to some power of the total number N of elements. As a consequence, the time of solution exponentially diverges for large N and the problem cannot be solved exactly even by the most powerful computers.

The NP-problems are not merely isolated exceptions, but are typical in the field of combinatorial optimization. Their existence imposes serious limitations on the performance of modern computers. On the other hand, the human brain does not experience any profound difficulty in dealing with the complex optimization problems that repeatedly arise in everyday life. Evidently, this property cannot contradict rigorous mathematical theorems, and hence it can never be guaranteed that the brain produces an *exact* solution with the best possible variant. In effect, the brain is able to find very quickly one of the *sufficiently good variants*.

Recently, several analog methods for the solution of combinatorial optimization problems were suggested that allow an approximate search for an optimal variant in the course of evolution of a certain nonlinear system. Often they make use of the idea of *simulated annealing* proposed by *Kirkpatrick* et al. [6.24]. We give an illustration of this approach for the *problem of graph bisection* [6.24] which is encountered, for instance, in the engineering of electronic chips. It is known that an exact solution of this problem requires an exponentially large number of operations.

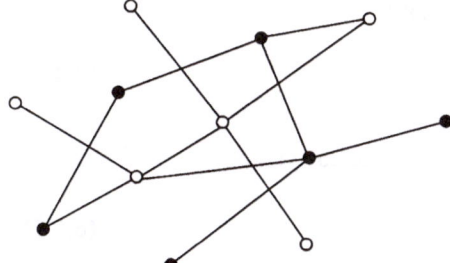

Fig. 6.3. Example of a graph bisection into two equal groups of elements (represented by white and black circles)

Suppose that a graph consists of an even number of sites and a set of connections between them. The optimization problem is to find such a division of all sites into two equal groups that the number of connections between the sites belonging to different groups is minimal (Fig. 6.3).

To formulate an analog method of solution, an artificial physical system should be constructed. We put into correspondence to every graph site a certain bistable physical element (i. e. an automaton) with two possible states specified by a binary spin variable S_i. Furthermore, we define the connection matrix $[T_{ij}]$ such that $T_{ij} = 1$ if the sites i and j are connected and $T_{ij} = 0$ if the connection is absent; for diagonal elements we choose $T_{ii} = 0$. Next we make a convention that all sites with $S_i = +1$ belong to the first group, while the sites with $S_i = -1$ belong to the second. Then the total number of connections N_b between the sites of these two groups can be written as

$$N_b = \frac{1}{8} \sum_{i,j} T_{ij} (S_i - S_j)^2 . \qquad (6.3.2)$$

Indeed, only if there is a connection between sites i and j (i.e. $T_{ij} = 1$) and they belong to different groups (i.e. $S_i = -S_j$), does this pair of sites give a contribution of four to the total sum; otherwise there is no contribution. Note also that every pair of elements is counted twice in (6.3.2).

Simple transformation of (6.3.2) yields

$$N_b = \frac{1}{2}N_0 - \frac{1}{2}\sum_{i,j} T_{ij}S_iS_j \, , \qquad (6.3.3)$$

where

$$N_0 = \sum_{i<j} T_{ij} \qquad (6.3.4)$$

is the total number of connections in the graph.

Note further that the "magnetization"

$$K_b = \sum_i S_i \qquad (6.3.5)$$

represents the difference in the numbers of sites in the two groups, and hence it is required to vanish for a valid solution. Therefore, the optimal bisection of the graph corresponds to an absolute minimum of the quantity

$$E = N_b + \lambda K_b^2 \qquad (6.3.6)$$

in the limit of large positive values of λ.

Substitution of expressions for N_b and K_b into (6.3.6) yields

$$E = \frac{1}{2}N_0 + \sum_{i,j}(\lambda - \frac{1}{4}T_{ij})S_iS_j \, . \qquad (6.3.7)$$

Now we see that E can be interpreted as the *energy* of interacting spins. Every spin configuration $\{S_i\}$ corresponds to a certain bisection of the graph. The ground state of such a spin system, for which energy E is minimal, corresponds to an optimal bisection[3].

Consequently, an analog solution of this optimization problem is possible if we choose such a dynamics, i.e. an algorithm for transitions between different states of the system which guarantee that the system relaxes to a state with an absolute minimum of energy.

At first glance, this can be achieved by using transition algorithm (6.2.5) of the Hopfield model. According to this algorithm an elementary spin-flip is allowed only if it leads to a lowering of the total energy. Thereby transitions continue until a state with a minimal energy is reached.

However, application of this steepest descent algorithm does not lead to a correct solution in the optimization problem. Actually, our system has a giant number of

---

[3] This is true in the limit $\lambda \to \infty$. When the parameter λ is large but finite, the numbers of elements in the two groups can be slightly different, which is usually quite acceptable in practical applications.

states corresponding to intermediate *local* energy minima. If an algorithm permits only such transitions that result in a lowering of energy, very soon the system becomes stuck in one of the local minima, which it is permanently unable to leave.

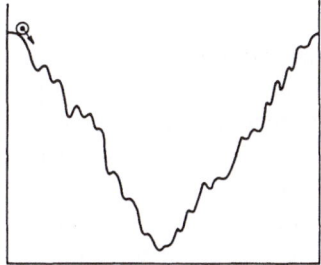

Fig. 6.4. Motion of a particle in a strongly jagged potential relief

This effect can be illustrated by the example of a particle that moves with large viscous friction in a strongly jagged potential relief (Fig. 6.4). Moving only down the potential gradient, this particle soon stops in one of the local minima and never reaches the bottom of the potential well.

A similar effect is well known in the physics of condensed matter. If a melt is quickly cooled, it usually hardens into an amorphous form characterized by irregular positioning of individual atoms. There are many such irregular structures, each corresponding to a certain metastable state with a local minimum of energy. The absolute minimum of energy corresponds, however, to a crystalline structure with periodic positioning of atoms. To bring the melt into a crystalline form, one should *anneal* it, i.e. cool the system in a very slow gradual way.

A similar *simulated annealing* technique can be used when we want to bring an artificial spin system into a configuration with an absolute minimum of energy.

Let us consider the following probabilistic algorithm for spin-flips:

If a flip $S_i \rightarrow S_i'$ results in a lowering of the total energy ($\Delta E \leq 0$) it is unconditionally accepted. However, if this flip leads to an increase of the total energy ($\Delta E > 0$) it can also be accepted, but with a probability

$$p(S_i \rightarrow S_i') = \exp(-\Delta E/\theta) \tag{6.3.8}$$

which is smaller the larger the increase of energy.

It can be shown [6.25] that such stochastic dynamics results asymptotically at large times in an equilibrium Boltzmann probability distribution with an effective temperature θ, for which the probability to find the system in a configuration with energy E is

$$P(E) = Z^{-1} \exp(-E/\theta) , \tag{6.3.9}$$

where Z is a normalization factor.

It is convenient to measure energy E from its absolute minimum, i.e. to put $E = 0$ there. Suppose that the initial spin configuration has high enough energy $E \gg \theta$. Then, at the beginning the spin system would steadily decrease its average energy because thermal fluctuations do not allow it to stop in configurations with local minima of energy. This descent would be continued until the system approaches configurations with energy E about θ. Thermal fluctuations are not desirable when the system is close to the deepest minimum because they prevent finding the optimal spin configuration which corresponds to the absolute minimum of energy. Indeed, as follows from (6.3.9), in a Boltzmann distribution all states with small energies $0 < E \ll \theta$ are realized with almost the same probability.

Simulated annealing means simply that we should slowly cool down the system (i.e. gradually decrease θ) at such a rate that the Boltzmann distribution is maintained. Then, in the final stage with vanishing temperature, thermal fluctuations are absent and a configuration with an absolute minimum of energy E is reached. This configuration defines the solution of the graph bisection problem.

Simulated annealing guarantees that we find an exact optimum only in the limit of infinitely slow cooling. If the total number of spins is very large, in the vicinity of an absolute energy minimum there might be many local minima with slightly higher energies. When we cool the system insufficiently slowly, it can remain in one of such intermediate states after all thermal fluctuations are switched off. However, this does not lead to serious mistakes because such configurations have almost the same energy as the best one and they can be used as approximate solutions of the optimization problem.

It is not always possible to find a physical system for which the optimization function would play the role of energy as was done above in the problem of graph bisection. Nevertheless, the technique of simulated annealing can be applied even in such cases. Next we show how this method can be used to obtain an approximate solution of the travelling salesman problem.

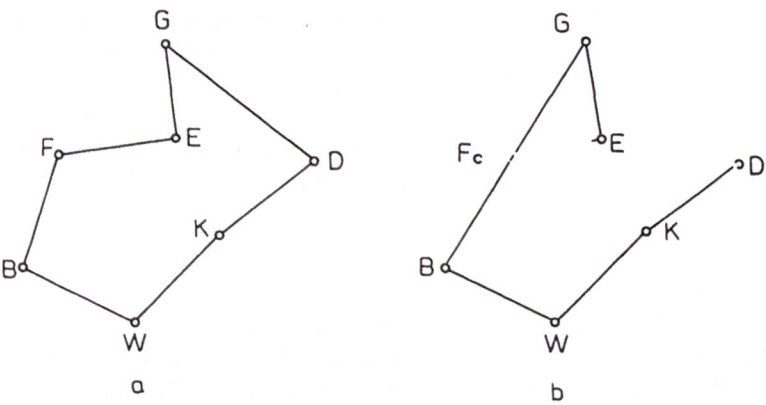

Fig. 6.5. Two tours (a,b) that differ in the order of visits to cities F and G

Let us assume that the tour is changed by mutations which each time involve only two arbitrarily chosen cities. When cities F and G are chosen as the candidates for a mutation at a given time step, the mutation consists in exchange of the positions occupied by these cities in the sequence of visits (Fig. 6.5). Thereby a new tour with a total length

$$l' = l_{BG} + l_{GE} + l_{EF} + l_{FD} + \ldots + l_{WB} \tag{6.3.10}$$

is constructed which differs from the length of the old tour (6.3.1) by

$$\Delta l = l_{BG} + l_{GE} + l_{EF} + l_{FD} - l_{BF} - l_{FE} - l_{EG} - l_{GD} . \tag{6.3.11}$$

Any tentative mutation is unconditionally accepted if it shortens the tour ($\Delta l < 0$) and accepted with a probability

$$p(\Gamma \rightarrow \Gamma') = \exp(-\Delta l / \theta) \tag{6.3.12}$$

if it increases the length of the tour ($\Delta l > 0$). When a mutation is accepted, the new order of visits $B, G, E, F, D, \ldots, W$ is taken as the initial one and the entire procedure is repeated. To simulate annealing, we slowly decrease parameter θ in (6.3.12), which plays a role of a temperature, and finally put it equal to zero. If the rate of cooling is sufficiently slow, this procedure allows us to find either the shortest possible tour or a good approximation.

Let us now discuss similarities and differences between complex combinatorial optimization and the problem of associative memory. When we solve an optimization problem, a typical situation is that there are many local optima but we are looking for a unique (absolute) optimum which is better than all others. Here, in order to find a correct solution, the outcome of the search procedure employed should not depend on initial conditions. To prevent the possibility of search termination at a local optimum, we introduce an analog of thermal fluctuations into the search algorithm.

On the other hand, in problems of associative memory the outcome *must* depend categorically on initial conditions. The search process should be terminated when the first fixed point is reached, even if it corresponds to one of the local minima of energy. Thus, associative memory and optimization represent two complementary classes of problems.

Introduction of thermal fluctuations is not obligatory for an analog solution of complex optimization problems. *Hopfield* and *Tank* [6.26] showed that deterministic analog techniques can be efficient.

The idea of Hopfield and Tank was to construct a dynamical system with continuous variables that would move along the direction of steepest descent to a state with a minimum of "energy" which defines the order of visits in the optimal tour for the travelling salesman problem. To obtain an analog solution of a problem with N cities, they had to use a nonlinear dynamical system with N^2 variables V_{xi}, each varying from 0 to 1, so that the permitted region of the phase space was the interior of a N^2-dimensional hypercube. All possible sequences of visits corresponded to different vertices of this hypercube. The dynamical equations were chosen in such a way that all the stable fixed points were just some of the hypercube vertices and every such fixed point corresponded to a certain permitted tour. The energy function reached its absolute minimum in those vertices of the N^2-dimensional hypercube that corresponded to the shortes tour. (Since any tour can be defined by $2N$ different sequences of visits, there are $2N$ different vertices for each tour.)

Every stable fixed point for the resulting dynamical system had its own basin of attraction and (in the absence of thermal fluctuations) the final state of the system was determined by its initial conditions. Hence, a special choice of initial conditions was made: they were taken almost at the center of the hypercube, so that no significant bias towards a particular vertex was created at the beginning.

Generally, deeper energy minima should possess larger attraction basins. Hopfield and Tank expected that the central region of the hypercube would belong to the attraction basins of those fixed points (i. e. the hypercube vertices) which corresponded to the absolute minimum of energy and determined the shortest tour.

Computer simulations [6.26] of this dynamic system confirmed that it is able to find tours which are either optimal or sufficiently good. For a 10-city problem Hopfield and Tank reported convergence in 50% of cases to one of the two shortest tours. Since the total number of variants for this problem is 10!/20 = 181440, this result seems rather good. In a problem with 30 cities the best tour found was 19% longer than the shortest.

A more effective deterministic method of analog solution for the travelling salesman problem was proposed by *Durbin* and *Willshaw* [6.27]. In this method a closed loop of elastic "string" is placed on the plane containing the cities which the salesman is to visit, and then slowly deformed until it forms a path connecting all the cities.

6.4 The Perceptron

The ability to learn constitutes an important feature of neural networks. We can start with a simple undifferentiated network and then *train* it to solve a particular class of tasks. In the process of training, the properties of individual elements (weights of connections and biases of neurons) are changed in a purposeful way, so that finally the network acquires the desired dynamics.

In principle, it is possible for an external instructor to calculate first all weights and biases, and afterwards to modify every single element in the required way. However, this is absolutely unrealistic for large networks with millions of connections. Hence, only such learning rules can be employed that do not involve detailed interference from the instructor. In the best case, the learning rule should be local and autonomous, i. e. any element of a network should be able to modify its own properties relying only on the information that comes from its neighbors.

As an illustration we can consider the Hebb learning rule (6.2.3). To implement it, one takes a network consisting of N binary neurons and $N(N-1)/2$ adjustable connections between them. Every connection is characterized by its own weight J_{ij}. The training process proceeds as follows. The instructor provides every neuron i with the value $\xi_i^{(m)}$ which should be taken by its spin variable S_i in a state corresponding to the mth recorded pattern. An automaton that realizes a connection between neurons i and j receives from them information about the respective values $\xi_i^{(m)}$ and $\xi_j^{(m)}$. It then modifies the value of the weight coefficient J_{ij}, stored in its memory, according to the rule

$$J'_{ij} = J_{ij} + (1/N)\xi_i^{(m)}\xi_j^{(m)} \ . \tag{6.4.1}$$

Afterwards new values $\xi_i^{(m+1)}$ are applied which correspond to the next recorded pattern, and the learning cycle is repeated. This procedure is continued until all the patterns are recorded. Initially we put $J_{ij} = 0$ for all connections.

When training is finished this network can be used to perform recognition of patterns. In such a regime all weight coefficients J_{ij} remain fixed. At the initial time moment, all elements are put into states ξ_i that correspond to the pattern which is presented for recognition. Later any external interference is switched off and neurons start to change their states according to the transition algorithm (6.2.5). After some time the transitions cease and the network comes into a stable state which corresponds to the nearest stored prototype pattern $\{\xi_i^{(m)}\}$. Instead of the Hebb learning rule, one can use local learning rules (6.2.14) or (6.2.15).

Classification performed by associative memory networks is very simple: each input pattern is classified according to its *overall* similarity with the stored prototypes, which is usually defined in terms of the overlaps $\sum_i \xi_i \xi_i^{(m)}$. This criterion is not satisfactory in many potential applications. Often the goal of learning is to teach the system to detect some *essential* features in the input patterns, not paying attention to irrelevant differences. Since the same essential feature might be present in patterns with very small total overlaps, systems with associative memory cannot be used in such situations.

Consider, for instance, the problem of *symmetry detection*. Here a system is required to divide all input patterns into two classes, depending on whether they are invariant under a certain symmetry transformation or not. Obviously, each class would include very different patterns with little overall similarity (i. e. with very small overlaps). Hence, it would not be possible to construct a single prototype for all patterns in the same class.

Generally, we want a system to perform categorization, based on a search for certain regularities in the input patterns.

The first attempt to construct such a system was made by *Rosenblatt* [6.28] who proposed a network known as the *Perceptron*. This network consists of two layers of input and output units (Fig. 6.6). Below we specify the states of input units by variables I_j and the states of output units by variables O_j; both variables are assumed to take only values 0 and 1. An input unit j is connected to an output unit i by a link with a weight J_{ij}. When the states of all input units are known, the states of the output units are determined by

$$O_i = H(a_i) , \tag{6.4.3}$$

where

$$a_i = \sum_j J_{ij} I_j , \tag{6.4.4}$$

and $H(z)$ is the step function, $H(z) = 0$ for $z \leq 0$ and $H(z) = 1$ for $z > 0$. Consequently, every activity pattern in the input layer produces a certain activity pattern of the output units. In other words, the system produces a definite *response* for every possible input (to perform classification, the system should produce identical responses for entire classes of different inputs).

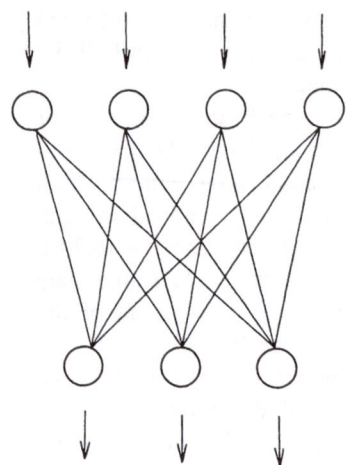

Fig. 6.6. The Perceptron network

Adjustment of responses is achieved by a training procedure which is accompanied by modifications of weights J_{ij}. In the process of training the system is presented with different input patterns $\{I_j\}$ for which the desired output patterns $\{D_i\}$ are known. Comparing every actual output pattern $\{O_i\}$ with the desired output $\{D_i\}$, it determines the errors

$$E_i = D_i - O_i \; . \tag{6.4.5}$$

When all errors are known, the system modifies weight J_{ij} by

$$\Delta J_{ij} = \varepsilon E_i I_j \; , \tag{6.4.6}$$

where ε is a small parameter that controls the rate of learning.

Since variables I_j, O_i and D_i take only values 0 and 1, the Perceptron learning algorithm (6.4.6) can be given a simple interpretation. The weights of connections leading to a given output unit i are changed only if its state is erroneous (i. e. it does not coincide with the desired output D_i). Moreover, such changes are performed only in the weights of connections that come from units j which are active for a presented pattern (i. e. if $I_j = 1$).

Training of a Perceptron consists in multiple presentations of different input patterns with prescribed outputs. After each new pattern is presented, all weights are corrected by using (6.4.6).

Instead of the deterministic response (6.4.3), probabilistic mapping is sometimes used. Then it is assumed that, for a given a_i, an output unit i is found in the active state ($O_i = 1$) with probability

$$p_i = \left[1 + \exp(-a_i/\theta)\right]^{-1} \; , \tag{6.4.7}$$

where θ plays the role of effective temperature. For $\theta = 0$ this probabilistic algorithm reduces to (6.4.3).

Recently the Perceptron scheme with probabilistic behavior ($\theta \neq 0$) was successfully used by *Rumelhart* and *McClelland* [6.29] in the linguistic problem of finding the past tenses of English verbs. In this computer experiment, the input consisted of various present tenses of verbs (encoded into binary sequences). The required response was the correct past tense of input verbs. As we know, English grammar is rather complex since it has many exceptional cases. Formation of past tenses of irregular verbs is governed by a complicated set of rules. Remarkably, a Perceptron system was able to learn these rules from experience achieved by training, without any interference by the human operator.

Nevertheless, the learning capacity of Perceptrons remain very limited. For instance, such a system cannot learn to reproduce the function of *exclusive or* (XOR). Let us consider this example in detail.

The logic function of XOR implies that an output unit (C) should go into an active state if only one of the two input units (A or B) is activated, i.e. the desired output is

$$D_C = \begin{cases} 0, & \text{if} \quad I_A = I_B = 0 , \\ 1, & \text{if} \quad I_A = 1 \quad \text{and} \quad I_B = 0 , \\ 1, & \text{if} \quad I_A = 0 \quad \text{and} \quad I_B = 1 , \\ 0, & \text{if} \quad I_A = I_B = 1 . \end{cases} \tag{6.4.8}$$

On the other hand, in a Perceptron with two input units A and B and one output unit C the actual output state O_C would be

$$O_C = H(J_A I_A + J_B I_B) , \tag{6.4.9}$$

or explicitly

$$O_C = \begin{cases} 0, & \text{if} \quad I_A = I_B = 0 , \\ H(J_A), & \text{if} \quad I_A = 1 \quad \text{and} \quad I_B = 0 , \\ H(J_B), & \text{if} \quad I_A = 0 \quad \text{and} \quad I_B = 1 , \\ H(J_A + J_B), & \text{if} \quad I_A = I_B = 1 . \end{cases} \tag{6.4.10}$$

To obtain $H(J_A) = H(J_B) = 1$ we need $J_A > 0$ and $J_B > 0$. But then $H(J_A + J_B) = 1$ and the output unit is in the active state when both input units are activated, which violates the definition of this logic function. Hence, weights J_A and J_B cannot be chosen in such a way that the function of "exclusive or" is realized.

Generally, a Perceptron can correctly classify only those patterns which are geometrically equivalent to regions in the vector space bounded by a plane (see *Minsky* and *Pappert* [6.30].

6.5 Back Propagation of Errors

To go beyond the serious limitations of the Perceptron, one can examine more complex systems that include several layers of *hidden* units placed between the input and the output layers. An effective learning algorithm for such a system was proposed by *Rumelhart* et al. [6.31, 32] (see also [6.33]).

Suppose that a learning system consists of N layers (Fig. 6.7). the first layer is made of input units that receive input signals $\{I_j\}$. The output signals $\{O_j\}$ are read

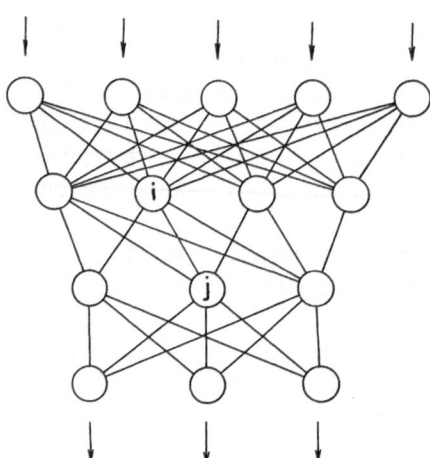

Fig. 6.7. Network with hidden units

from the units of the last layer. Between these two layers there are $(N-2)$ layers of hidden units. Only the units which belong to neighboring layers are connected. The total number of units can differ from one layer to another.

When unit i in layer n receives signal $x_i^{(n)}$, it produces an output signal[4]

$$y_i^{(n)} = \left[1 + \exp(-x_i^{(n)})\right]^{-1} . \tag{6.5.1}$$

In its turn, the input signal received by unit j in the next $(n+1)$-th layer is formed as a weighted sum of the output signals from the previous nth layer, i.e.

$$x_j^{(n+1)} = \sum_i J_{ji}^{(n)} y_i^{(n)} . \tag{6.5.2}$$

Weights $J_{ji}^{(n)}$ can take both positive and negative values.

The input for the units of the first layer is simply the pattern presented for analysis, i.e. $x_j^{(1)} = I_j$. The output of the last layer units is the result of processing, $y_j^{(N)} = O_j$.

If the weights of connections are random, the response of the system, i.e. the output pattern, will be random as well. The purpose of learning is to adjust the weights so that the desired responses are produced.

Specifically, the problem is formulated as follows. Suppose we want a system to classify all possible input patterns into several distinct categories according to a certain criterion. Then we form a training set that includes sufficiently large numbers of input patterns $\{I_{j,m}\}$ corresponding to different categories and indicate a desired category output $\{D_{j,m}\}$ for each of them. During the process of learning, our system should adjust its weights $J_{ji}^{(n)}$ in such a manner that it responds by producing the desired output every time the corresponding training pattern is presented. Afterwards, we can take a new pattern, not included in the training set, and show it to the system.

---

[4] The particular form of the function $y = f(x)$ is not very essential. It is only necessary that it monotonically increases from 0 to 1 when x varies from $-\infty$ to $+\infty$. However, if the function (6.5.1) is used, this simplifies the analysis.

If the training set was sufficiently large and representative, we expect that the system will be able to produce a correct response to this new pattern, i.e. to perform its categorization.

Let us define the error function as

$$E = \frac{1}{2} \sum_{m=1}^{M} \sum_{j} (D_{j,m} - O_{j,m})^2 , \tag{6.5.3}$$

where $O_{j,m} = y_{j,m}^{(N)}$ are the actual outputs of the last layer units when the mth training pattern $\{I_{j,m}\}$ is presented to the units of the first layer. Obviously, this function reaches its absolute minimum only if for all M training patterns actual outputs coincide with the desired outputs. Our aim is to find the values of the weights $J_{j,m}^{(n)}$ which minimize (6.5.3). Hence, we see that learning is a complex optimization problem.

To minimize E by using a steepest descent method, we should calculate partial derivatives of E with respect to the weights of every connection. Any such derivative is given by a sum of quantities related to different training patterns. For a given training pattern, these derivatives of the error function can be calculated explicitly by employing the technique of back propagation of errors which was introduced by *Rumelhart* et al. [6.31–32] and by *Le Cun* [6.33].

Note first of all that the identity

$$\frac{\partial E}{\partial y_i^{(n-1)}} = \sum_{j} \frac{\partial E}{\partial y_j^{(n)}} \frac{\partial y_j^{(n)}}{\partial x_j^{(n)}} \frac{\partial x_j^{(n)}}{\partial y_i^{(n-1)}} , \tag{6.5.4}$$

holds. Here we assume that a certain training pattern m is applied to the first layer, so that $x_j^{(n)}$ and $y_j^{(n)}$ are input and output signals corresponding to this particular pattern, i.e. $y_j^{(n)} = y_{j,m}^{(n)}$.

By using (6.5.1) and (6.5.2), we obtain from (6.5.4) a recurrent formula

$$\frac{\partial E}{\partial y_i^{(n-1)}} = \sum_{j} \frac{\partial E}{\partial y_j^{(n)}} y_j^{(n)} (1 - y_j^{(n)}) J_{ji}^{(n-1)} . \tag{6.5.5}$$

This allows us to calculate the derivatives $\partial E / \partial y$ for a given layer of units if we know the values of the derivatives for the next layer. Note that for the last layer these derivatives can be easily found by taking a derivative of (6.5.3):

$$\frac{\partial E}{\partial y_i^{(N)}} = y_i^{(N)} - D_i . \tag{6.5.6}$$

Hence, by starting from the lowest Nth layer and moving upwards, we can subsequently calculate by (6.5.5) all partial derivatives $\partial E / \partial y$ for every unit.

Furthermore we can use the identity

$$\frac{\partial E}{\partial J_{ji}^{(n-1)}} = \frac{\partial E}{\partial y_j^{(n)}} \frac{\partial y_j^{(n)}}{\partial x_j^{(n)}} \frac{\partial x_j^{(n)}}{\partial J_{ji}^{(n-1)}} , \tag{6.5.7}$$

which can be written as

$$\frac{\partial E}{\partial J_{ji}^{(n-1)}} = \frac{\partial E}{\partial y_j^{(n)}} y_j^{(n)} \left(1 - y_j^{(n)}\right) y_i^{(n-1)} , \qquad (6.5.8)$$

if we take into account (6.5.1) and (6.5.2).

When all derivatives $\partial E / \partial y$ are calculated, (6.5.8) allows us to find the weight derivatives $\partial E / \partial J$ for every connection at a fixed training pattern.

The values of derivatives $\partial E / \partial J$ thus found can be used further to correct the weights $J_{ji}^{(n)}$ after every subsequent training pattern is presented. An alternative approach would be to present in turn, in a given learning cycle, *all* M training patterns, to sum values of derivatives $\partial E / \partial J$ for all presented patterns and only after this to insert corrections into the weights. Whatever the particular procedure employed, every learning cycle includes presentation of training patterns, calculation of derivatives of the error function E with respect to the weights of all connections and a small modification of every weight by a correction

$$\Delta J_{ji}^{(n)} = -\varepsilon \frac{\partial E}{\partial J_{ji}^{(n)}} , \qquad (6.5.9)$$

after which the entire learning cycle is repeated.

Learning should be continued until an acceptably small value of the error (6.5.3) is reached. If the learning procedure does not allow us to significantly reduce the errors, it can be performed again with an increased number of hidden units. Instead of (6.5.9) it is better to use a correction algorithm

$$\Delta J_{ji}^{(n)}(t) = -\varepsilon \frac{\partial E}{\partial J_{ji}^{(n)}} + \kappa \Delta J_{ji}^{(n)}(t-1) , \qquad (6.5.10)$$

where discrete time t increases by 1 after every learning cycle, and coefficient κ ranging from 0 to 1 specifies inertiality of the learning system.

Greater learning abilities arise if the system units have permanent biases, so that (6.5.2) is replaced by

$$x_j^{(n+1)} = \sum_j J_{ji}^{(n)} y_i^{(n)} + b_j^{(n)} . \qquad (6.5.11)$$

The values $b_j^{(n)}$ of biases can be also adjusted in the process of learning. Note that introduction of biases does not lead to any essential changes in the above arguments. Indeed, the presence of a permanent bias for a given unit is formally equivalent to receiving a fixed input signal $Y_j^{(n)} = 1$ through an additional connection with the weight $b_j^{(n)}$. Hence, biases can be considered on the same footing as weight coefficients $J_{ji}^{(n)}$. Similarly to (6.5.8) we find

$$\frac{\partial E}{\partial b_j^{(n-1)}} = \frac{\partial E}{\partial y_j^{(n)}} y_j^{(n)} \left(1 - y_j^{(n)}\right) . \qquad (6.5.12)$$

When the derivatives are known, biases can be corrected by using, for instance,

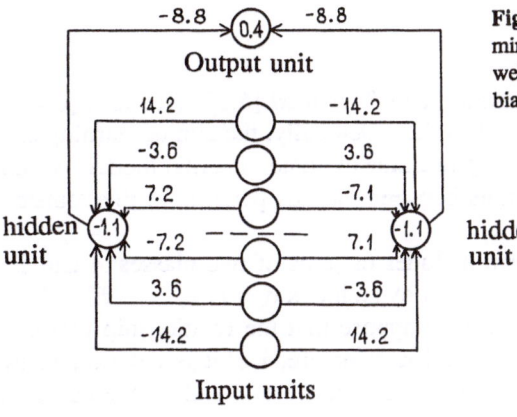

Fig. 6.8. A network that has learned to detect mirror symmetry. The numbers on the arcs are weights and the numbers inside the nodes are biases

$$\Delta b_j^{(n)} = -\varepsilon \frac{\partial E}{\partial b_j^{(n)}} . \qquad (6.5.13)$$

A simple task which cannot be performed by connecting input units directly to output units (i.e. by means of a Perceptron) is the detection of mirror symmetry. To detect whether an activity pattern in a one-dimensional array of input units is symmetric about the centre point, it is essential to use an intermediate layer of hidden units. In a computer simulation by *Rumelhart* et al. [6.31, 34], employing back propagation of errors, an elegant solution of this problem was found which involved only two hidden units (Fig. 6.8). The system was trained with 64 patterns. Every learning cycle consisted in presentation of all these patterns followed by calculation of derivatives $\partial E/\partial J$ and $\partial E/\partial b$ for every unit. Then these derivatives were summed over all training patterns; the results were used to modify the weights and biases according to (6.5.10). After that the entire learning cycle was repeated. Initial weights were randomly chosen in the interval from -0.3 to 0.3. Parameters ε and κ in (6.5.10) had values $\varepsilon = 0.1$ and $\kappa = 0.9$. After 1425 learning cycles the system achieved the ability to detect presence of mirror symmetry in *any* one-dimensional activity pattern (this can easily be checked by using the values of weights and biases indicated in Fig. 6.8).

Other examples of learning by back propagation of errors are given in [6.35–38].

The technique of back propagation of errors employs a deterministic optimization algorithm to search for the required connection weights. Therefore, there is always a danger that, by moving strictly down the gradient of the error function E, we reach one of the local minima of this function and the process of learning will be terminated despite the fact that no genuine solution is found. If this happens, one can try to add new layers of hidden units. This effectively increases the dimensionality of the space of weight coefficients. Then one can hope that some new valleys will emerge which will destroy the barriers separating bad local minima in a space of lower dimensionality. Nevertheless, no general proof of convergence for back propagation of errors is at present available and it is doubtful that it exists at all.

6.6 The Boltzmann Machine

The *Boltzmann machine* proposed by *Hinton* and *Sejnowski* [6.39] (see also [6.40]) is the most universal system with analog learning. Generally, the aim of learning can be viewed as the construction by an analog machine of some "internal model" which can reproduce regularities in the relations between various patterns in the "outside world".

Let us assume that the outside world includes patterns of two classes A and B; we enumerate the patterns belonging to class A by an index $\alpha = 1, 2, \ldots, K$ and the patterns from class B by $\beta = 1, 2, \ldots, M$. Suppose that the relationship between the patterns of these two classes is of a statistical nature, i. e. it is determined by joint probabilities $p(\beta, \alpha)$ of observation of various patterns β and α. We can also define conditional probabilities

$$\pi(\beta|\alpha) = \frac{p(\beta, \alpha)}{p(\alpha)} \tag{6.6.1}$$

which tell us what is the probability to observe pattern β if another pattern α is present in the environment (here $p(\alpha)$ is the probability to observe α).

Suppose further that any pattern α corresponds to a certain input of the analog machine, while different patterns β are associated with its possible responses. Then we can say that this machine has a valid internal model if the statistical relationship between its inputs and responses is the same as the relationship between the corresponding patterns in the outside world. Hence, learning can be interpreted as a process of gradual construction of an internal model.

The Boltzmann machine represents a network of bistable units, each with two states specified by a binary variable s_1 that takes values 0 and 1. The units are not grouped into layers; instead any unit is symmetrically connected to all the others. For any activity pattern $\{s_j\}$ of this network, energy E is defined as

$$E = -\frac{1}{2} \sum_{i,j} J_{ij} s_j s_j + \sum_i b_i s_i , \tag{6.6.2}$$

where J_{ij} are the weights of connections ($J_{ii} = 0$) and b_i are the biases.

All units in the Boltzmann machine are divided into *visible* (input and output units) and *hidden*. External patterns are applied to input units, and corresponding responses are displayed by output units.

Application of a particular pattern clamps input units, but leaves free all other units of the network. Free units undergo transitions in accordance with the probabilistic algorithm which is specified below. Note that any permanent bias b_i in (6.6.2) can be always realized by a connection between unit i and an additional input unit which is constantly clamped in the active state; then b_i is the weight of such a connection. Therefore, from the theoretical point of view it is sufficient to consider a network without biases, i. e. with energy

$$E = -\frac{1}{2} \sum_{i,j} J_{ij} s_i s_j; . \tag{6.6.3}$$

When unit i goes from the state of rest into the active state, energy E is changed by an amount

$$\Delta E_i = \sum_j J_{ij} s_j \ . \tag{6.6.4}$$

The probabilistic transition algorithm is such that, for given states of all other units, unit i goes into the active state ($s_i = 1$) at the next time step with a probability

$$p_i = \left[1 + \exp(\Delta E_i / \theta) \right]^{-1} \ , \tag{6.6.5}$$

where θ plays the role of a temperature.

By standard methods of statistical physics it can be shown that such probabilistic transitions finally bring this network to "thermal equilibrium" where the probability to observe any particular pattern $\{s_i\}$ of the network activity is given by the *Boltzmann distribution*

$$P(\{s_i\}) = Z^{-1} \exp \left[-E(\{s_i\}) / \theta \right] \ . \tag{6.6.6}$$

Here Z is a normalization factor.

As we have mentioned, for a Boltzmann machine the goal of learning consists in construction of the "internal model" which would reproduce the probabilistic relationships between patterns of classes A and B in the outside world. This implies that various patterns α and β of these two classes should be coded first in some way into the corresponding binary patttterns of activity $\boldsymbol{I}^{(\alpha)} = (I_i^{(\alpha)}, \ \dots \ , I_n^{(\alpha)})$ and $O^{(\beta)} = (O_1^{(\beta)}, \ \dots \ , O_{n'}^{(\beta)})$ of input and output units.

Every "internal model" represents a particular set of connection weights J_{ij} in the Boltzmann machine. When all these weights are fixed and input units are clamped in the states corresponding to a certain activity pattern $\boldsymbol{I}^{(\alpha)}$, the network will generate with some probability different activity patterns $O^{(\beta)}$ of output units. Watching the behavior of this system, we can find the conditional probabilities $\pi'(\beta|\alpha)$ to find an output pattern β if some pattern α is applied to input units. The process of learning consists in a gradual change of weights J_{ij} which brings closer the conditional probabilities $\pi'(\beta|\alpha)$ generated by the "internal model" and the conditional probabilities $\pi(\beta|\alpha)$ existing in the outside world.

As the criterion of a fit between two probability distributions we can take the value of the so-called *information theoretic measure*

$$G = \sum_{\alpha, \beta} p(\beta, \alpha) \ln \frac{\pi(\beta|\alpha)}{\pi'(\beta|\alpha)} \ . \tag{6.6.7}$$

It can be shown that G reaches its absolute minimum $G = 0$ when probabilities π and π' coincide for all pairs (β, α).

Hence, the problem is reduced to minimization of G. Let us find the derivatives $\partial G / \partial J_{ij}$. Note first that weights of connections influence only the conditional probability π' in (6.6.7) and therefore

$$\frac{\partial G}{\partial J_{ij}} = - \sum_{\alpha, \beta} p(\beta, \alpha) \frac{\partial}{\partial J_{ij}} \ln \pi'(\beta|\alpha) \ . \tag{6.6.8}$$

Hence, we have to calculate the logarithmic derivate of π' with respect to J_{ij} for given activity patterns I and O of input and output units.

Let us denote by $S = \{s_i\}$ activity patterns of hidden units. Then, when some input pattern I is fixed, the probability fo find the subsystems of hidden and output units in the states S and O, respectively, is given by the expression

$$P_I(S, O) = Z_I^{-1} \exp\left[-E_I(S, O)/\theta\right] ,$$
(6.6.9)

where Z_I is defined as

$$Z_I = \sum_{S,O} \exp\left[-E_I(S, O)/\theta\right] .$$
(6.6.10)

Since the conditional probability $\pi'(O|I)$ is related to $P_I(S, O)$ by

$$\pi'(O|I) = \sum_S P_I(S, O) ,$$
(6.6.11)

we have

$$\pi'(O|I) = Z_I^{-1} \sum_S \exp\left[-E_I(S, O)/\theta\right] .$$
(6.6.12)

Introducing a quantity

$$Z_{I,O} = \sum_S \exp\left[-E_I(S, O)/\theta\right]$$
(6.6.13)

which is calculated under clamped states of *both* the input and the output units, we can rewrite (6.6.12) as

$$\pi'(O|I) = Z_{I,O}/Z_I .$$
(6.6.14)

Consequently,

$$\frac{\partial}{\partial J_{ij}} \ln \pi'(O|I) = Z_{I,O}^{-1} \frac{\partial Z_{I,O}}{\partial J_{ij}} - Z_I^{-1} \frac{\partial Z_I}{\partial J_{ij}} .$$
(6.6.15)

If we take into account the definitions of Z_I and $Z_{I,O}$, as well as the dependence of energy E on weights of connections (cf. (6.6.3)), we find from (6.6.15) that

$$\frac{\partial}{\partial J_{ij}} \ln \pi'(O|I) = (\langle s_i s_j \rangle_{I,O} - \langle s_i s_j \rangle_I)/\theta ,$$
(6.6.16)

where $\langle s_i s_j \rangle_I$ and $\langle s_i s_j \rangle_{I,O}$ are the time averages determined under the condition of the clamped states of only the input units (I) or both the input and the output (O) units, respectively.

Substitution of (6.6.16) into (6.6.9) yields

$$\frac{\partial G}{\partial J_{ij}} = \frac{1}{\theta} \sum_{\alpha,\beta} p(\beta, \alpha)(\langle s_i s_j \rangle_\alpha - \langle s_i s_j \rangle_{\alpha,\beta}) .$$
(6.6.17)

When the derivatives $\partial G/\partial J_{ij}$ are known, they can be used to modify the weights J_{ij} in order to decrease G. This can be done, for instance, by making the corrections[5]

$$\Delta J_{ij} = -\varepsilon \frac{\partial G}{\partial J_{ij}} \, . \tag{6.6.18}$$

In an important special case when every input pattern α is uniquely associated with a definite desired response $\beta = \beta(\alpha)$ and all patterns α are equally probable, equation (6.6.17) takes the form

$$\frac{\partial G}{\partial J_{ij}} = \frac{1}{K\theta} \sum_{\alpha} (\langle s_i s_j \rangle_\alpha - \langle s_i s_j \rangle_{\alpha,\beta(\alpha)}) \, , \tag{6.6.19}$$

where K is the total number of different input patterns α.

First we give the description of the learning procedure in this special case. Every learning cycle consists of the following steps:

A. Input units are clamped in the states that correspond to an input pattern α, whereas output units are clamped in the states imposed by the corresponding output pattern $\beta(\alpha)$. Then the network is brought to thermal equilibrium and, by averaging over long time intervals, correlations $\langle s_i s_j \rangle_{\alpha,\beta(\alpha)}$ are calculated for all pairs (i, j). This step is repeated for all input patterns α from the training set.

B. Input units are again clamped in the pattern α, but output units are left free. After the network is brought to thermal equilibrium, correlations $\langle s_i s_j \rangle_\alpha$ should be calculated for all pairs. This step is also repeated for all patterns α.

C. Calculated correlations $\langle s_i s_j \rangle_{\alpha,\beta(\alpha)}$ and $\langle s_i s_j \rangle_\alpha$ are then used to determine the derivative (6.6.19) and to modify the connection weights using (6.6.18).

Such learning cycles should be repeated until an acceptable frequency of correct associations is reached. If there is no good convergence, the entire procedure can be repeated with an increased number of hidden units.

Despite its apparent simplicity, this learning algorithm has one subtle point. If we want to construct a Boltzmann machine capable of deterministic association, it requires a low temperature θ (otherwise statistical fluctuations will be large and no clear association will be realized). When the temperature is low, at thermal equilibrium the network will mostly be found in the activity patterns that have the lowest energy (6.6.3). But besides a central deepest energy minimum the network will surely possess many other local energy minima separated by sufficiently large barriers. In the process of relaxation this system can come to such local minima and wait there until the barrier is overcome due to a rare fluctuation. The waiting time diverges exponentially for vanishing temperature. Hence, if we start with a small temperature, it would take an unacceptably long time to bring the network to the thermal equilibrium which is assumed by expressions (6.6.19).

This difficulty can be removed by the application of simulated annealing (see Sect. 6.3). At every step A or B we should start at a sufficiently high temperature

---

[5] Other variants of the steepest descent can be used, as well. For example, in [6.41] the rule $\Delta J_{ij} = -\varepsilon \, \text{sign} \, (\partial G/\partial J_{ij})$ was chosen.

and then gradually decrease it so that the network permanently remains in thermal equilibrium. Note that θ in (6.6.19) is the final temperature at which annealing ends.

A similar procedure can be used to learn probabilistic associations. As we have mentioned, in this case any input pattern α is allowed to produce different responses β, with relative frequencies determined by conditional probability (6.6.1). Hence, at step A we clamp input and output units in *arbitrarily* chosen patterns α and β and, after simulated annealing, determine various correlations $\langle s_i s_j \rangle_{\alpha,\beta}$ at thermal equilibrium for a given set of weights J_{ij}. Step B remains unchanged. At step C, however, (6.6.17) is used instead of (6.6.19) to calculate the derivatives $\partial G / \partial J_{ij}$.

Note that the learning procedure can also be modified so that all weights J_{ij} are modified by

$$\Delta J_{ij} = (\varepsilon/\theta) \sum_{\alpha,\beta} p(\beta,\alpha)\langle s_i s_j \rangle_{\alpha,\beta} \tag{6.6.20}$$

after every step A and by

$$\Delta J_{ij} = -(\varepsilon/\theta) \sum_{\alpha} p(\alpha)\langle s_i s_j \rangle_{\alpha} \tag{6.6.21}$$

after every step B.

Thus, weights J_{ij} are incremented following step A and decremented after step B. We can say that positive learning occurs in the first phase during which information from the environment is imprinted in the weights of connections. The second phase represents *unlearning* during which the system randomly samples states according to their Boltzmann distribution.

A good qualitative explanation of unlearning was suggested by *Hinton* and *Sejnowski* [6.40]:

"Consider a hidden unit deep within the network: How should its connections with other units be changed to best capture regularity present in the environment? If it does not receive direct input from the invironment, the hidden unit has no way to determine whether the information it receives from neighboring units is ultimately caused by structure in the environment or is entirely a result of the other weights. This can lead to a *folie à deux* where two parts of the network each construct a model of the other and ignore the external environment. The contribution of internal and external sources can be separated by comparing co-occurences in *phase*[+] [i. e. in step A] with the similar information that is collected in the absence of environmental input. *Phase*[−] [i. e. step B] thus acts as a control condition. Because of the special properties of equilibrium it is possible to subtract off this purely internal contribution and use the difference to update weights. Thus, the role of the two phases is to make the system maximally responsive to regularities present in the environment and to prevent the system from using its capacity to model internally-generated regularities."

Various applications of the Boltzmann machine are discussed in [6.34, 41–43]. Its major disadvantage consists in a very long learning time which is necessary because we have to perform annealing and to collect statistics for many pairs of patterns. This is the price that should be paid for universality. Good knowledge can be acquired only by slow learning.

6.7 Storage of Temporal Sequences

Above we discussed associative memory and learning for static patterns. However, the repertoire of analog information processing by neural networks (natural or artificial) is not entirely limited to operations with static patterns. There are many tasks which require storage and associative recall of *temporal sequences*. The human memory stores not only pictures, but also melodies and poems which can be retrieved by presentation of their short or distorted fragments. Another important function of the brain is generation of rhythmic motor patterns which control, for instance, swimming or locomotion. Clearly, the same abilities are desirable in artificial systems of analog information processing.

Several closely related approaches to the problem of storage of temporal sequences were proposed by *Amari* [6.44], *Kleinfeld* [6.45], *Personnaz* et al. [6.46], and *Sompolinsky* and *Kanter* [6.47].

Suppose we have an ordered set of patterns $\{\xi_i^\mu\}$, $\mu = 1, 2, \ldots, q$, and want to construct a neural network which will generate a periodic sequence of patterns

$$\{\xi_i^1\}, \{\xi_i^2\}, \ldots, \{\xi_i^\mu\}, \{\xi_i^{\mu+1}\}, \ldots, \{\xi_i^q\}, \{\xi_i^1\}, \ldots . \tag{6.7.1}$$

For simplicity we assume now that all these patterns are orthogonal, i. e. condition (6.2.2) holds.

The storage problem is most easily solved (see [6.46]) for the networks with parallel synchronous updating (see Sect. 6.1):

$$S_i(t + 1) = \text{sign} \left[\sum_j J_{ij} S_j(t) \right] . \tag{6.7.2}$$

Then it is sufficient to choose (asymmetric) synaptic matrix J_{ij} in the form

$$J_{ij} = \frac{1}{N} \sum_{\mu=1}^q \xi_i^{\mu+1} \xi_j^\mu , \tag{6.7.3}$$

where we assume $\xi_i^{q+1} = \xi_i^1$.

Suppose we start at $t = 1$ with the first pattern, i. e. $S_i(1) = \xi_i^1$. Then we find

$$\begin{aligned}
\sum_j J_{ij} \xi_j^1 &= \frac{1}{N} \sum_j \sum_\mu \xi_i^{\mu+1} \xi_j^\mu \xi_j^1 \\
&= \sum_\mu \xi_i^{\mu+1} \delta_{\mu 1} = \xi_i^2
\end{aligned} \tag{6.7.4}$$

and, as follows from (6.7.2),

$$S_i(2) = \xi_i^2 . \tag{6.7.5}$$

At the next time step $t = 3$ this network will generate the third pattern, and so on until the qth step after which it generates again the first pattern and the cycle is repeated.

In networks with sequential updating this simple procedure does not work. Since the states of individual neurons are then updated asynchronously, the network quickly develops a mixture of two patterns $\{\xi_i^\mu\}$ and $\{\xi_i^{\mu+1}\}$ and afterwards its performance becomes completely smeared.

To stabilize alternating patterns we can introduce *delays* into the sequential updating algorithm and add a symmetric component into the synaptic matrix, as suggested by *Sompolinski* and *Kanter* [6.47] and by *Herz* et al. [6.48]. Let us consider a sequential updating algorithm

$$S_i'(t) = \text{sign } h_i(t) \tag{6.7.6}$$

where effective field h_i consists of two contributions, $h_i(t) = h_i^{(1)}(t) + h_i^{(2)}(t)$, given by

$$h_i^{(1)}(t) = \sum_j J_{ij}^{(1)} S_j(t) , \tag{6.7.7}$$

$$h_i^{(2)}(t) = \sum_j J_{ij}^{(2)} \int_{-\infty}^{t} w_0(t - t') S_j(t') \, dt' . \tag{6.7.8}$$

Here $h_i^{(1)}$ describes synchronous interactions between the spins that are characterized by a symmetric matrix

$$J_{ij}^{(1)} = \frac{1}{N} \sum_{\mu=1}^{q} \xi_i^\mu \xi_j^\mu . \tag{6.7.9}$$

The second component $h_i^{(2)}$ includes delays, so that the value of $h_i^{(2)}$ at a given time moment is determined by the spin configurations that existed at earlier times t'. This delayed interaction has an asymmetric matrix

$$J_{ij}^{(2)} = (\lambda_0/N) \sum_{\mu=1}^{q} \xi_i^{\mu+1} \xi_j^\mu , \tag{6.7.10}$$

which differs from (6.7.3) by a constant factor λ_0. We assume that the delay function $w_0(t)$ in (6.7.8) falls to zero within a time τ (which is much longer than the average time required to update in turn all the spins in the system) and that

$$\int_0^{+\infty} w_0(t) \, dt = 1 . \tag{6.7.11}$$

First we discuss the special case of a fixed delay, $w_0(t) = \delta(t - \tau)$. Then $h_i^{(2)}(t)$ depends only on the spin configuration at the time moment $t - \tau$. Suppose that the network stayed until time τ in a particular pattern $\{\xi_i^\mu\}$. What will be its state at the next time moment? This is determined by the sign of the total field h_i at $t > \tau$. Expressions (6.7.7–10) yield

$$h_i(t) = \xi_i^\mu + \lambda_0 \xi_i^{\mu+1} . \tag{6.7.12}$$

Since binary variables ξ_j take only the values ± 1, the last term in (6.7.12) cannot change the sign of h_i if $\lambda_0 < 1$. In this case the network will continue to stay in the old state $\{\xi_i^\mu\}$.

On the other hand, if $\lambda_0 > 1$ the sign of h_i is determined by the second term in (6.7.12), which implies that the network will experience a transition to a new pattern $\{\xi_i^{\mu+1}\}$ at time moment $t = \tau$. Repeating the same arguments, we can show that at time $t = 2\tau$ this network will go to the next pattern $\{\xi_i^{\mu+2}\}$ and so on. Hence, when $\lambda_0 > 1$ such network will indeed generate the periodic sequence of patterns (6.7.1), with every pattern remaining for a short time $t_0 = \tau$ before being replaced by the next pattern.

Note that in this special case of a fixed delay we can omit the first (synchronous) contribution (6.7.7) to the effective field h_i. Formally, this corresponds to the limit $\lambda_0 \to \infty$. However, this contribution plays an important role for more complex delay functions. For instance, in the case of the exponential delay function $w_0(t) = \tau^{-1} \exp(-t/\tau)$, *Sompolinsky* and *Kanter* [6.47] found that the time t_0 during which each pattern is stable is given by the expression

$$t_0 = \tau \left\{ \ln 2 - \ln \left[1 - (2/\lambda_0 - 1)^{1/2} \right] \right\} . \tag{6.7.13}$$

Simple examination shows that this result is meaningful only for $1 < \lambda_0 < 2$.

For a general delay function $w_0(t)$ the duration t_0 of a single pattern can be estimated (see [6.47]) from equation

$$\int_{t_0}^{2t_0} w_0(t)\, dt = \frac{1}{2}(1 - 1/\lambda_0) . \tag{6.7.14}$$

We assumed above that all patterns $\{\xi_i^\mu\}$ are orthogonal. Obviously, this condition is too restrictive. In effect, the same procedure can be used (see [6.47]) for random unbiased patterns, provided that their number is small compared with the total number of spins.

There is also another method, proposed by *Personnaz* et al. [6.46, 48] (see also [6.49]), which allows work with correlated and biased patterns. It represents a modification of the pseudo-inverse rule (6.2.12) which is used in models of static associative memory. For an arbitrary set of patterns $\{\xi_i^\mu\}$, matrices (6.7.9) and (6.7.10) should be replaced by

$$J_{ij}^{(1)} = \frac{1}{N} \sum_{\mu=1}^{q} \xi_i^\mu \chi_j^\mu , \tag{6.7.15}$$

$$J_{ij}^{(2)} = (\lambda_0/N) \sum_{\mu=1}^{q} \xi_i^{\mu+1} \chi_j^\mu , \tag{6.7.16}$$

where

$$\chi_i^\mu = \sum_{\nu=1}^{q} (R^{-1})_{\mu\nu} \xi_i^\nu \tag{6.7.17}$$

and R^{-1} is the inverse to the overlap matrix

$$R_{\mu\nu} = \frac{1}{N} \sum_i \xi_i^\mu \xi_i^\nu . \tag{6.7.18}$$

It can easily be verified that all previous arguments remain valid when matrices (6.7.15) and (6.7.16) are used instead of $J_{ij}^{(1)}$ and $J_{ij}^{(2)}$. Note that the same replacement can be performed in the case of synchronous parallel updating.

To conclude this section we would like to discuss yet another approach to the storage of temporal sequences, which was proposed by *Buchmann* and *Schulten* [6.50]. It is formulated for a network of binary units with two possible states $s_i = 0$ and $s_i = 1$. Probabilistic updating is assumed: each unit can go at the next time moment to the active state $s_i = 1$ with the probability

$$p_i = \left\{ 1 + \exp\left[-\left(\sum_j J_{ij} s_j - B \right) \Big/ \theta \right] \right\}^{-1} . \tag{6.7.19}$$

where B is a certain threshold.

This procedure can be used for storage of low activity patterns $\{s_i^\mu\}$, $\mu = 1, 2, \ldots, q$, that have only a small fraction of active units. Furthermore, for these patterns the orthogonality condition

$$\sum_i \varepsilon^\mu s_i^\mu s_i^\nu = \delta_{\mu\nu} \tag{6.7.20}$$

holds, where ε^μ is the inverse of the number of active units in a pattern μ.

We begin with construction of a symmetric matrix J_{ij} that allows storage of $\{s_i^\mu\}$ as stable steady patterns. This can be achieved if we take this matrix in the form

$$J_{ij} = \sum_{\mu=1}^q \left(\varepsilon^\mu s_i^\mu s_j^\mu - \sum_{\substack{\nu=1 \\ \nu \neq \mu}}^q \varkappa_1 (q/N) s_i^\mu s_j^\nu \right) . \tag{6.7.21}$$

The first (activatory) term here induces a cooperation between neurons that represent the same pattern μ. The second (inhibitory) term gives rise to competition between all stored patterns. It can be shown that the resulting dynamics asymptotically selects a network state which coincides with the pattern which has the largest overlap with the initial state, i. e. this network can serve as a system of associative memory.

In order to retrieve the patterns $\{s_i^\mu\}$ in the sequence $\mu = 1, 2, \ldots, q$, we modify the symmetric synaptic matrix J_{ij} given by (6.7.21). Two principal alterations are made. We eliminate the inhibition between predecessor-successor patterns in the sequence, i. e. between $\{s_i^\mu\}$ and $\{s_i^{\mu+1}\}$. Moreover, we add an excitatory (positive) projection from $\{s_i^\mu\}$ to $\{s_i^{\mu+1}\}$ and an inhibitory backward-projection from $\{s_i^{\mu+1}\}$ to $\{s_i^\mu\}$. After such changes, the synaptic matrix becomes

$$J_{ij} = \sum_{\mu=1}^{q} \left(\varepsilon^{\mu} s_i^{\mu} s_j^{\mu} - \sum_{\substack{\nu=1 \\ |\nu-\mu|>1}}^{q} \varkappa_1 (q/N) s_i^{\mu} s_j^{\nu} + \varkappa_2 \varepsilon^{\mu-1} s_i^{\mu-1} s_j^{\mu} - \varkappa_3 \varepsilon^{\mu+1} s_i^{\mu+1} s_j^{\mu} \right) .$$

(6.7.22)

Buhman and *Schulten* [6.50] performed computer simulations for such a network. In their experiment, the network retrieved a sequence of six patterns representing numbers 1, 2, ... , 6.

6.8 Networks with Mediators

The distinctive feature of neural network models is their high connectivity. For instance, in the Hopfield model of associative memory every element should be linked to all the others. Since the total number of required connections increases as the square of the total number of elements, this presents a serious difficulty in attempts to implement such networks on a microscopic scale, as molecular electronic devices. From the point of view of such technology, it is preferable to use networks that only have local regular connections.

In this section we show that many neural network models can be reformulated in local terms, with interactions between units transmitted by *mediators*.

Let us first consider a probabilistic Hopfield model with N spins S_i. Suppose that, in one time unit, any spin can experience a flip $S_i \rightarrow S_i'$ with a probability $W(S_i \rightarrow S_i')$ which is determined by the momentary spin configuration. Namely, we assume

$$W(S_i \rightarrow S_i') = w \exp\{h_i(S_i' - S_i)/\theta\} ,$$

(6.8.1)

where

$$h_i = \sum_j J_{ij} S_j .$$

(6.8.2)

When temperature θ goes to zero, only transitions that decrease the total energy are allowed. The overall rate of spin-flips is specified by parameter w. When the Hebbian learning rule is used we have

$$J_{ij} = \frac{1}{N} \sum_{\mu=1}^{M} \xi_i^{\mu} \xi_j^{\mu} .$$

(6.8.3)

The corresponding local model with mediators is constructed as follows. Suppose that the "spins" (which actually represent some two-state units) are immersed in the medium, where certain particles which we call *mediators* are diffused. There are many types of mediators, each corresponding to a particular stored pattern $\{\xi_i^{\mu}\}$. For simplicity, we will assume that the diffusion of mediators is sufficiently fast to ensure ideal mixing; hence, no spatial distribution effects will be finally taken into account.

Any spin, located at a site R_i, generates mediators of type μ when $S_i = \xi_i^\mu$, i.e. when its current direction coincides with the direction of this spin for the μth stored pattern. For a fixed spin configuration $\{S_i\}$, the local generation rate ϱ_μ of mediators μ at point r is

$$\varrho_\mu(r) = \gamma(V/N) \sum_i (\xi_i^\mu S_i + 1)\delta(r - R_i) , \qquad (6.8.4)$$

where V is the volume of the medium. All mediators are subjected to decay at a constant rate γ (this prevents their accumulation in the medium).

Mediators of each type μ act upon the spins, forcing them to come into the state defined by the corresponding μth stored pattern. The local rates of induced spin-flips are given by

$$\Phi_i(S_i \to S_i') = w \exp\left[\theta^{-1} \sum_{mu} \xi_i^\mu (n_\mu(R_i) - 1)(S_i' - S_i)\right] . \qquad (6.8.5)$$

where n_μ is the local density of mediators μ.

By comparing (6.8.5) and (6.8.1) we see that spins experience thermal flips in an effective field

$$h_i = \sum_\mu \xi_i^\mu (n_\mu(R_i) - 1) \qquad (6.8.6)$$

which is determined by momentary population densities of mediators.

The kinetic equations for the population densities n_μ of mediators are simply

$$\dot{n}_\mu = -\gamma n_\mu + \gamma(V/N) \sum_i (\xi_i^\mu S_i + 1)\delta(r - R_i) + D\Delta n_\mu . \qquad (6.8.7)$$

where D is the diffusion constant of the mediator.

If diffusion is fast enough to guarantee ideal mixing, uniform distribution of mediators is maintained. If, furthermore, the lifetimes of mediators are much shorter than the intervals between successive spin-flips (i.e. if $\gamma \gg w$) uniform population densities of mediators adiabatically adjust to the current spin configuration $\{S_i\}$, so that

$$n_\mu = \frac{1}{N} \sum_i (\xi_i^\mu S_i + 1) . \qquad (6.8.8)$$

Substituting (6.8.8) into (6.8.5) we find

$$\Phi_i(S_i \to S_i') = w \exp\left[\frac{1}{\theta N} \sum_j \sum_\mu \xi_i^\mu \xi_j^\mu S_j(S_i' - S_i)\right] . \qquad (6.8.9)$$

This expression coincides with the rate (6.8.1) of spin-flips in the probabilistic Hopfield model with the Hebbian learning rule. Hence, in the limit $\gamma \gg w$ the model with mediators reduces to the standard Hopfield model.

It is interesting to consider the opposite limit $\gamma \ll w$ of extremely fast spin-flips. In this case spins adjust adiabatically to the momentary population densities of mediators that determine the effective field (6.8.6). The Boltzmann probability distribution for a spin S_i in a given field h_i at temperature θ is

$$p(S_i) = \frac{\exp(h_i S_i/\theta)}{\exp(-h_i/\theta) + \exp(h_i/\theta)} \ . \tag{6.8.10}$$

Since characteristic times of mediators are very long, mediators feel only time-averaged values of spin variables which coincide with the statistical averages $< S_i >$ estimated with the Boltzmann distribution (6.8.10), i.e.

$$\langle S_i \rangle = \sum_{S_i = \pm 1} S_i p(S_i) = \tanh(h_i/\theta) \ . \tag{6.8.11}$$

Substituting $\langle S_i \rangle$ instead of S_i into (6.8.7) and taking into account (6.8.6) we find the kinetic equations for population densities of mediators

$$\dot{n}_\mu = -\gamma n_\mu + \gamma (V/N) \sum_i \left\{ \xi_i^\mu \tanh \left[\theta^{-1} \sum_\nu \xi_i^\nu (n_\nu(\mathbf{r}) - 1) \right] + 1 \right\} \delta(\mathbf{r} - \mathbf{R}_i)$$
$$+ D\Delta n_\mu \ . \tag{6.8.12}$$

Thus, we have derived a system of multi-component reaction-diffusion equations that describe the evolution of population densities of mediators associated with different stored patterns. Now we can discuss their solutions in several important cases.

If we introduce new variables

$$m_\mu = n_\mu - 1 \ , \tag{6.8.13}$$

equations (6.8.12) take the form

$$\dot{m}_\mu = -\gamma m_\mu + \gamma (V/N) \sum_i \xi_i^\mu \tanh \left(\theta^{-1} \sum_\nu \xi_i^\nu m_\nu(\mathbf{r}) \right) \delta(\mathbf{r} - \mathbf{R}_i)$$
$$+ \gamma \left[(V/N) \sum_i \delta(\mathbf{r} - \mathbf{R}_i) - 1 \right] + D\Delta m_\mu \ . \tag{6.8.14}$$

Suppose now that diffusion is very fast and mediators are uniformly distributed in the medium. Then (6.8.14) yields

$$\dot{m}_\mu = -\gamma m_\mu + (\gamma/N) \sum_i \xi_i^\mu \tanh \left(\theta^{-1} \sum_\nu \xi_i^\nu m_\nu \right) \ . \tag{6.8.15}$$

At low temperatures θ this set of differential equations has many attractive fixed points. In this limit, $\tanh(x/\theta) \approx \text{sign } x$ and (6.8.15) can be written as

$$\dot{m}_\mu = -\gamma m_\mu + (\gamma/N) \sum_i \xi_i^\mu \text{ sign} \left(\sum_\nu \xi_i^\nu m_\nu \right) \ . \tag{6.8.16}$$

Below we consider a case when all stored patterns $\{\xi_i^\mu\}$ are orthogonal (see (6.2.2)). In this case (6.8.16) has simple stationary solutions for which $m_\mu = \delta_{\mu\alpha}$ with $\alpha = 1, 2, \ldots, M$. Such solutions are stable with respect to small perturbations. Indeed, for $m_\mu = \delta_{\mu\alpha} + m_\mu'$ we find from (6.8.16):

$$
\begin{aligned}
\dot{m}_\mu' &= -\gamma m_\mu' - \gamma \delta_{\mu\alpha} + (\gamma/N) \sum_i \xi_i^\mu \xi_i^\alpha \\
&= -\gamma m_\mu' .
\end{aligned}
\tag{6.8.17}
$$

These attractive fixed points have a very simple interpretation. At equilibrium the equality

$$
m_\mu = \frac{1}{N} \sum_i \xi_i^\mu \langle S_i \rangle
\tag{6.8.18}
$$

holds and, therefore, the m_μ represent *overlaps* between the established pattern $\{S_i\}$ and different stored patterns $\{\xi_i^\mu\}$. When one of the overlaps is equal to unity, $\{S_i\}$ coincides with the corresponding stored pattern. Hence, fixed point $m_\mu = \delta_{\mu\alpha}$ corresponds to the state with the αth stored pattern. Depending on the initial conditions, the evolution of this system ends in one of the attractive fixed points. Hence, in the limit $\gamma \ll w$ our system with mediators has the same dynamical properties as the Hopfield model.

In effect, the m_μ can be considered as *order parameters* of the Hopfield model. The dynamical equations (6.8.15) for these parameters were derived by *Rieger* et al. [6.51], without appealing to the notion of mediators. Similar dynamical equations with delays were constructed in [6.48] for a related problem of temporal sequence storage.

The above analysis shows that in both limiting cases $\gamma \gg w$ and $\gamma \ll w$ the local model with mediators is effectively equivalent to the Hopfield model of associative memory if diffusion is fast enough to maintain ideal mixing within the medium volume V, i.e. if $D/\gamma \gg V^{1/3}$. More sophisticated examination (see [6.52]) shows that this equivalence in the asymptotic dynamical properties is valid for an arbitrary relationship between the characteristic rates γ and w.

Moreover, the local model with mediators can easily be generalized to other learning rules which allow the expression of matrix $[J_{ij}]$ in the form

$$
J_{ij} = \sum_\mu \varphi_i^\mu \Psi_j^\mu ,
\tag{6.8.19}
$$

where $\{\varphi_i^\mu\}$ and $\{\Psi_j^\mu\}$ are pairs of associated patterns. Then ξ_i^μ should be replaced by Ψ_i^μ in (6.8.4) and by φ_i^μ in (6.8.5).

When matrix $[J_{ij}]$ cannot be decomposed as (6.8.9), another approach [6.53] using the concept of mediators may be applied. In this case mediators should be *addressable*, i.e. they should bear a "tag" where the address of the corresponding spin is indicated.

To summarize the discussion in this section, our analysis reveals an intimate relationship between neural networks and active media described by multicomponent reaction-diffusion equations.

7. Reproductive Networks

In this chapter we consider multicomponent active systems of reproductive agents. Such agents are assumed to interact, either directly or through some mediators, so that the interactions influence their reproduction rates. If a particular interaction enhances the reproduction rate, it is activatory; if it reduces the rate of reproduction, it can be considered as inhibitory. Such reproductive networks comprise a large number of models that are found in various fields, from chain nuclear reactions to prebiological evolution, ecological systems, and market economies.

In the simplest case, when all agents compete for a common renewable resource and no directed interactions are present, competition leads to survival of the single fittest agent. If, in addition to the common resource, each agent has its own specialized resource, their steady coexistence is possible. On the other hand, when any agent inhibits reproduction of all the others, the system becomes multistable, i.e. it has many steady states that correspond to survival of different agents. In the networks with directed interactions, reproduction waves can propagate.

Since reproductive networks are capable of very complex dynamic behavior, they can be used for the purposes of analog information processing (it can be also argued that some natural reproductive networks are actually performing this function). In this aspect, evolution turns out to be closely related to learning.

7.1 Competition and Cooperation of Reproductive Agents

Below we formulate several basic models of reproductive networks. To provide illustrations, we use greatly simplified examples from prebiological evolution, evolutionary ecology, and market economics.

Let us consider a system of N biological species characterized by their populations n_1, $i = 1, 2, \ldots, N$. The populations are numerous and therefore the variables n_i can be treated as continuous. Population dynamics is described by a set of differential equations

$$\dot{n}_i = (K_i - \gamma_i)n_i , \quad i = 1, 2, \ldots, N , \tag{7.1.1}$$

where K_i is the mean individual reproduction rate and γ_i is the mean death rate of species i.

Both reproduction and death rates can depend on populations of other species and on concentrations c_k of some important resources (food, oxygen, etc.) in the environment, i.e.

$$K_i = K_i \left(\{n_i\}, \{c_k\} \right), \ \gamma_i = \gamma_i \left(\{n_i\}, \{c_k\} \right) . \tag{7.1.2}$$

First we consider the situation when there are no interactions between species, except for the competition for a common resource (say, a certain kind of food). It can be assumed that reproduction rates K_i are proportional to the concentration c of food, $K_i = a_i c$, and death rates are constant. Then (7.1.1) takes the form

$$\dot{n}_i = (a_i c - \gamma_i) n_i , \ i = 1, 2, \ldots , N . \tag{7.1.3}$$

The amount of food available depends, generally, on the populations n_i of consumers.

Suppose food grows at some constant rate S and decays at a rate Γ. Individuals of different species i consume food at a rate proportional to its average density in the medium, i.e. at a rate $r_i c$. Then we have

$$\dot{c} = S - \left(\sum_i r_i n_i + \Gamma \right) c . \tag{7.1.4}$$

Suppose food density c changes so quickly that it adjusts adiabatically to the momentary populations n_i. Then (7.1.4) yields

$$c = S / \left(\sum_i r_i n_i + \Gamma \right) . \tag{7.1.5}$$

Provided that populations n_i are sufficiently small, (7.1.5) can be approximated by

$$c = \left[S - \sum_i r_i (S/\Gamma) n_i \right] / \Gamma . \tag{7.1.6}$$

Introducing new notations $u_i = (b_i/\Gamma) n_i$, $k_i = S/\Gamma - \gamma_i/a_i$, $\alpha_i = a_i$, we obtain after substitution of (7.1.6) into (7.1.4) the typical equations

$$\dot{u}_i = \alpha_i \left(k_i - \sum_j u_j \right) u_i , \tag{7.1.7}$$

which describe competition of species relying entirely on the same resource (kind of food).

In a more complicated situation when, besides one common resource, there are also some separate resources which are associated with particular species, (7.1.7) is replaced by

$$\dot{u}_i = \alpha_i \left(k_i - \beta_i u_i - \sum_{j \neq i} u_j \right) u_i , \tag{7.1.8}$$

where $\beta_i > 1$.

Until now it was assumed that death rates γ_i are not influenced by existing populations. However, it is also possible that each species produces "poisons" which inhibit the reproduction of other species, i. e. increase their death rates.

When inhibition is nonspecific, any species equally increases the death rates of all other species. If this influence is small enough, death rates depend linearly on the populations, i. e.

$$\gamma_i = \gamma_i^0 + g \sum_{j \neq i} n_j . \tag{7.1.9}$$

The effect of nonspecific inhibition can be included in the previous model. It does not change the form of equations (7.1.8). However, now the coefficients β_i can also be smaller than 1 if mutual inhibition is sufficiently strong.

Besides nonspecific competition, various directed interactions between species are possible. For instance, the product of species i might serve as food or an essential resource for another species j. Then it *activates* reproduction of this species j. It is also possible that species i produces some poison which selectively *inhibits* reproduction of a certain species j. Such interactions can be taken into account by introducing additional terms into (7.7.8), which yields

$$\dot{u}_i = \alpha_i \left(k_i - \beta_i u_i - \sum_{j \neq i} u_j + \sum_{j \neq i} \kappa_{ij} u_j - \sum_{j \neq i} \mu_{ij} u_j \right) . u_i . \tag{7.1.10}$$

Here κ_{ij} and u_{ij} specify the degree of activatory and inhibitory influence by the jth species on the ith species, respectively. When there is no activatory or inhibitory interaction between two given species, these coefficients vanish. Since interactions are directed, matrices $[\kappa_{ij}]$ and $[\mu_{ij}]$ are generally not symmetric.

If we define the *interaction matrix* $[J_{ij}]$ as

$$J_{ij} = \kappa_{ij} - \mu_{ij} - 1 , \quad i \neq j , \tag{7.1.11}$$

we can write (7.1.10) in the form

$$\dot{u}_i = \alpha_i \left(k_i - \beta_i u_i - \sum_{j \neq i} J_{ij} u_j \right) u_i . \tag{7.1.12}$$

Although (7.1.10) and (7.1.12) were derived for a system of interacting biological species, these equations are found in other situations. *Eigen* and *Schuster* [7.1, 2] showed that similar equations describe the networks of catalytic chemical reactions which play a fundamental role in prebiological evolution. In the latter case, however, they should be completed by an additional term that takes into account *mutations*[1].

---

[1] The biochemical kinetic equations usually include also the terms of higher order in u_i, but this difference is not principal.

If a "species" represents some kind of replicating molecule, mutations can be viewed as rare random events that transform one such "species" into another. Assuming that w_{ij} is the rate of mutation of species j into species i, we find

$$\dot{u}_i = \alpha_i \left(k_i - \beta_i u_i - \sum_{j \neq i} J_{ij} u_j \right) u_i + \sum_{j \neq i} w_{ij} u_j - \sum_{j \neq i} w_{ji} u_i . \tag{7.1.13}$$

The first additional term in (7.1.13) describes an increase in the population of species i due to mutations that transform into it the other species. The last term takes into account losses of the population of species i due to mutations into the other species.

Quite remarkably, the same or very similar mathematical models describe phenomena at the opposite extreme of the vastly different levels of the structural hierarchy leading from molecules to social systems. As an illustration, we consider below a greatly simplified mathematical model of market economics.

Suppose we have a group of agents, labelled by index i, that produce a certain kind of commodity and compete on the market. Every agent produces per unit time an amount q_i of this commodity (hence, q_i can be called the *production* of a given agent). The net cost to an individual agent of the produced commodity is $V_i = E_i q_i$, where E_i is its specific cost which includes expenditures for raw materials, machine depreciation, labour payments, etc. For simplicity, we can include into E_i also a minimal, generally established, profit rate needed for technological innovations and financial insurance.

When a commodity comes to market, it is sold at a price E which is determined by market conditions. If the market equilibrium is established (i. e. the equilibrium between the demand and the supply of goods is reached), the market price E represents a weighted average of individual costs, i. e.

$$E = \frac{\sum_i E_i q_i}{\sum_i q_i} . \tag{7.1.14}$$

This assumption is quite natural: if some agent produces a commodity in very small amounts, it cannot seriously influence the market price, whatever its individual net cost of production.

The surplus profit G_i gained by the agent i is

$$G_i = E q_i - E_i q_i . \tag{7.1.15}$$

A part of this profit is further invested to expand production. Hence, the rate of production expansion dq_i/dt is proportional to the gained profit G_i, i. e.

$$\dot{q}_i = \alpha (E - E_i) q_i , \tag{7.1.16}$$

where α is a proportionality factor.

Equations (7.1.16) and (7.1.14) describe, in the simplest approximation, the economic competition of agents in a free market. They are very similar to the model that describes competition of biological species. The analogy between market economics and biological evolution was noticed a long time ago (see [7.3, 4]). Evolutionary approaches in studies of market economics and in the related field of the spread of technological innovation were discussed recently by *Nelson* and *Winter* [7.5], *Ebeling* and *Jimenez Montano* [7.6], and *Silverberg* [7.7].

Actually, economic agents not only compete but also cooperate one with another. Consider, for instance, a reproductive network where each agent makes fixed payments, proportional to his, her, or its production, to other members of the reproductive network. Such payments can be described by adding two new terms to equation (7.1.16), so that it becomes

$$\dot{q}_i = \alpha(E - E_i)q_i + \sum_j w_{ij}q_j - \sum_j w_{ji}q_i \,. \qquad (7.1.17)$$

The first term takes into account the flow of payments from other agents to the agent i; the second term describes the reverse flow from ith agent to all other agents. Positive coefficients w_{ij} characterize the degree of cooperation between agents i and j.

Certainly, there are many other forms of cooperation (and inhibition) between economic agents. Some of them can lead to the same directed interactions which were discussed above in the evolutionary context.

In a slightly different form, equations (7.1.17) also arise in evolutionary biological models with "constant organization". These and further examples can be found in the review of molecular evolutionary models by *Schuster* [7.8] and in the book by *Ebeling* and *Feistel* [7.9].

7.2 Selection

We begin our analysis of reproductive networks with the simplest case described by models (7.1.7) or (7.1.16). In this special case there is only one kind of interaction: all reproductive agents compete for the same resource (food, investments, etc.). A general consequence of such competition is *selection of the fittest*. It turns out that, in the long time limit, only the agent with the best performance is left to operate, whereas all other less effective agents are eliminated.

First we consider the model (7.1.16) which allows a complete analytical solution. It is convenient to introduce new variables

$$y_i = \frac{q_i}{\sum_l q_l} \,, \qquad (7.2.1)$$

which specify the relative contribution of ith agent to the total production $Q = \sum_l q_l$.

Note that, according to (7.1.16), the total production remains fixed. Indeed, we have

$$\dot{Q} = \alpha E \sum_i q_i - \alpha \sum_i E_i q_i \,, \qquad (7.2.2)$$

but the right side of this equation vanishes because of the definition (7.1.14) of E. Hence, $Q = \text{const}$.

In terms of these new variables, equations (7.1.16) take the form

$$\dot{y}_i = \alpha(E - E_i)y_i \tag{7.2.3}$$

where

$$E = \sum_l E_l y_l . \tag{7.2.4}$$

Next we go to variables Y_i defined by

$$Y_i = y_i \exp\left(-\alpha \int_0^t E(t')dt'\right) . \tag{7.2.5}$$

As follows from (7.2.3), these variables satisfy linear equations

$$\dot{Y}_i = -\alpha E_i Y_i \tag{7.2.6}$$

which can be easily solved, yielding

$$Y_i(t) = Y_i(0) \exp(-\alpha E_i t) . \tag{7.2.7}$$

Since $\sum_i y_i = 1$, (7.2.5) gives

$$\sum_l Y_l = \exp\left(-\alpha \int_0^t E(t')dt'\right) \tag{7.2.8}$$

and therefore (using again (7.2.5) and (7.2.7)) we find

$$y_i(t) = \frac{y_i(0) \exp(-\alpha E_i t)}{\sum_l y_l(0) \exp(-\alpha E_l t)} . \tag{7.2.9}$$

Thus we have obtained a general solution of (7.2.3) and (7.1.16) for arbitrary initial productions q_i (or $y_i = q_i/Q$). Now we can show that this solution indeed describes "survival of the fittest".

Suppose that the production cost E_1 of the first agent is minimal, i.e. $E_1 = \min E_l$. Then it makes a contribution to the denominator of (7.2.9) which declines least rapidly with time, and hence dominates at sufficiently large times. At $t \to \infty$ we find

$$y_i(t) \approx [y_i(0)/y_1(0)] \exp[\alpha(E_1 - E_i)t] . \tag{7.2.10}$$

Therefore, production of all agents, except for the most effective first one, vanishes with time and these agents are finally eliminated.

Similar behavior is found for model (7.1.7). The equations of this model can be written as

$$\frac{d}{dt}\alpha_i^{-1} \ln u_i = k_i - \sum_l u_l . \tag{7.2.11}$$

Therefore, for any two species i and j we have

$$\frac{d}{dt}[\alpha_i^{-1}\ln u_i - \alpha_j^{-1}\ln u_j] = k_i - k_j .\tag{7.2.12}$$

This implies

$$\frac{u_i^{1/\alpha_i}}{u_j^{1/\alpha_j}} = \text{const} * \exp\left[(k_i - k_j)t\right] .\tag{7.2.13}$$

Hence, we see that the ratio of two populations goes to zero in the limit $t \to \infty$ if $k_i < k_j$.

Suppose that the first species is most efficient, i. e. k_1 is maximal. Then (7.2.13) shows that all other populations decline exponentially with time compared to the population of the first species (which approaches value $u_1 = 1$ as $t \to \infty$). Hence, model (7.1.7) describes selection of the most efficient reproductive agent (i. e. of the fittest biological species).

It was assumed above that there are no specific interactions between the competing agents. Next we consider the case when, in addition to nonspecific competition for the common resource, some of these agents directly influence reproduction of the others. Namely, we discuss below the competition of *mutating* reproductive agents.

In molecular biology, mutations represent physical changes of one replicating molecule into another, as described by equations (7.1.13). However, very similar mathematical models arise also on other applications where the meaning of "mutations" is different. For example, similar behavior is found for an economic community with mutual financial assistance of agents (i. e. with partial redistribution of profits).

Consider equations (7.1.17) that were derived in Sect. 7.1 for the above mentioned economic model but also describe mutations in populations of replicating molecules (see [7.1,2,9]). After transformations (7.2.1) and (7.2.5) they take the form

$$\dot{Y}_i = -\alpha E_i Y_i + \sum_l w_{il} Y_l - \sum_l w_{li} Y_i .\tag{7.2.14}$$

If we introduce formal vectors $Y = \{Y_i\}$ these equations can be written as

$$\dot{Y} = -AY ,\tag{7.2.15}$$

where $A = B + W$, and matrices $B = [B_{ij}]$ and $W = [W_{ij}]$ are defined by

$$B_{ij} = \alpha E_i \delta_{ij} ,\tag{7.2.16}$$

$$W_{ij} = -w_{ij} + \sum_l w_{li} \delta_{ij} .\tag{7.2.17}$$

For simplicity let us assume that mutation rates w_{ij} are symmetric, i. e. $w_{ij} = w_{ji}$. Then matrix A is also symmetric and it has a set of linearly independent eigenvectors Y^σ with real eigenvalues λ_σ:

$$AY^\sigma = \lambda_\sigma Y^\sigma .\tag{7.2.18}$$

Any vector Y can be decomposed as

$$Y = \sum_\sigma C^\sigma Y^\sigma , \tag{7.2.19}$$

where C^σ are suitable coefficients. If we substitute this decomposition into (7.2.14) we find that the C^σ should obey equations

$$\dot{C}^\sigma = -\lambda_\sigma C^\sigma . \tag{7.2.20}$$

Hence, the general solution of (7.2.14) is

$$Y(t) = \sum_\sigma C^\sigma(0) \exp(-\lambda_\sigma t) Y^\sigma . \tag{7.2.21}$$

According to (7.2.5) we have

$$y_i = Y_i \exp\left(\alpha \int_0^t E(t')\, dt'\right) . \tag{7.2.22}$$

It can easily be verified that equations (7.1.17) conserve total production Q and therefore $\sum_i y_i = 1$. If we sum over i in (7.2.22) this again yields (7.2.8). Consequently, in terms of variables y_i the general solution reads

$$y_i(t) = \frac{\sum_\sigma C_\sigma(0) \exp(-\lambda_\sigma t) Y_i^\sigma}{\sum_l \sum_\sigma C_\sigma(0) \exp(-\lambda_\sigma t) Y_l^\sigma} \tag{7.2.23}$$

where the Y_i^σ are the components of Y^σ.

Suppose that λ_1 is the smallest of all eigenvalues λ_σ. Then the corresponding exponent dominates both in the nominator and the denominator of (7.2.23), and in the limit $t \to \infty$ we have

$$y_i \approx \frac{Y_i^1}{\sum_l Y_l^1} . \tag{7.2.24}$$

We see that mutations lead to an important change in the outcome of evolution. In biological terms, instead of selection of a single fittest species a certain asymptotic distribution of species populations is formed. The relative weights of different populations for this distribution are determined by the components Y_i^1 of the eigenvector Y^1 that coresponds to the smallest eigenvalué λ_1 of the matrix A.

Moreover, by comparing (7.2.23) and (7.2.9) one can conclude that each population distribution which corresponds to some eigenvalue λ_σ behaves as if it were a single species with $E_\sigma = \lambda_\sigma$. As noted by *Eigen* [7.1], mutations lead to the formation of *quasispecies* that play effectively the same role as individual species in the absence of mutations.

In the models of market economics, "mutations" are equivalent to payment of financial aid between certain agents. The above discussion shows that this should result in the formation of stable conglomerates of agents which enter the competition as single entities and are analogous to quasispecies.

When the environment is fluctuating (so that coefficients E_i change in time) conglomerates have higher chances of survival than individual "egotistic" agents. It is very probable that an agent which may become "the fittest" after some future environment change is already present in an existing conglomerate, even when this agent is not yet the most efficient.

Usually, the rates of mutations w_{ij} are small (in economic terms this means that the size of the distributed aid is small compared to the total income). This allows us to treat matrix W in (7.2.14) as a perturbation. By using the standard perturbation theory for eigenvalue problems we find in the first approximation (see [7.10])

$$y_i^1 \approx \frac{w_{1i}}{E_i - E_1} , \quad i \neq 1 , \tag{7.2.25}$$

$$y_1^1 = 1 - \sum_{i \neq 1} y_i^1 . \tag{7.2.26}$$

Therefore, the quasispecies consists mainly of the individuals of the strongest (first) species. The fraction of other species is proportional to the rate of mutation.

7.3 Coexistence and Multistability

When every reproductive agent relies entirely upon its own resource, which might represent a particular kind of food for biological species or a particular group of consumers for the goods produced by a firm, all these agents can coexist with each other. The situation becomes more complicated if, in addition, such agents have also some common resource for which they compete. Similar behavior is found when, instead of a common resource, there is some nonspecific inhibition between agents.

Such situations are described by model (7.1.8). Below we consider for simplicity only a special case with equal coefficients $\beta_i = \beta$ which already illustrates the basic phenomena. In this case the model is given by equations

$$\dot{u}_i = \alpha_i \left(k_i - \beta u_i - \sum_{j \neq i} u_j \right) u_i . \tag{7.3.1}$$

When coefficient β is equal to unity, all agents use the same resource; this case was already discussed in Sect. 7.2. On the other hand, if $\beta \gg 1$ the dominant role in saturation of the population growth of any species is played by exhaustion of its own resource, and competition for a common resource is not essential. As noted in Sect. 7.1.1, model (7.3.1) describes also nonspecific inhibition of agents relying on separate resources. In the latter case β can be either larger or smaller than unity; the *smaller* β is, the stronger the mutual inhibition.

For convenience, let us enumerate all N reproductive agents in such a way that

$$k_1 > k_2 > \ldots > k_i > \ldots > k_N . \tag{7.3.2}$$

Furthermore, we will assume in the first part of our analysis that $\beta > 1$.

Suppose we have a steady coexisting group that consists of the first m species. Their respective stationary populations should satisfy equations

$$k_i - (\beta - 1)u_i - U = 0 , \quad i = 1, 2, \ldots, m , \tag{7.3.3}$$

where

$$U = \sum_{j=1}^{m} u_j \; . \tag{7.3.4}$$

The solution to (7.3.3) is

$$u_i^0 = (\beta - 1)^{-1}(\beta + m - 1)^{-1} \left[(\beta - 1 + m)k_i - \sum_{j=1}^{m} k_j \right] \; . \tag{7.3.5}$$

Since populations u_i^0 cannot be negative, this solution is meaningful only if

$$k_i \geq (\beta - 1 + m)^{-1} \sum_{j=1}^{m} k_j \quad \text{for any} \quad i \leq m \; . \tag{7.3.6}$$

If we introduce Λ_m by

$$\Lambda_m = \sum_{j=1}^{m} (k_j/k_{m+1} - 1), \quad \Lambda_0 = 0 \; , \tag{7.3.7}$$

and take into account that k_m is the smallest of the k_i, condition (7.3.6) can be formulated as

$$(\beta - 1) \geqslant \Lambda_{m-1} \; . \tag{7.3.8}$$

This imposes a lower limit on β for the steady coexistence of m species.

In order to correspond to a steady coexistence, the stationary solution (7.3.5) should be stable, at least under small perturbations. One should distinguish between the *internal* and the *external* stability of the coexisting group.

External stability means that, in presence of m coexisting species, the populations of all other $(N - m)$ species with smaller efficiencies k_i should be suppressed. This implies the conditions

$$k_i - \sum_{j=1}^{m} u_j^0 < 0 \quad \text{for} \quad m < i \leq N \; . \tag{7.3.9}$$

Clearly, it is sufficient that (7.3.9) is satisfied for the largest of the involved k_i, i.e. for k_{m+1}. Substitution of (7.3.5) into (7.3.9) then yields

$$\Lambda_m > (\beta - 1) \; . \tag{7.3.10}$$

Internal stability means that the stationary solution (7.3.5) should be stable in respect to small variations δu_i in populations of m coexisting species. Linearizing (7.3.1) with respect to such small variations, we find

$$\delta \dot{u}_i = -\alpha_i u_i^0 [(\beta - 1)\delta u_i + \delta U], \quad i = 1, 2 \ldots, m \; , \tag{7.3.11}$$

where

$$\delta U = \sum_{j=1}^{m} \delta u_j \, . \tag{7.3.12}$$

This system of m linked linear differential equations has solutions of the form

$$\delta u_i(t) \sim \exp(-\lambda t) \, . \tag{7.3.13}$$

By substitution of (7.3.13) into (7.3.11) it can be found that all values of λ are given by the roots of the algebraic equation

$$1 = \sum_{i=1}^{m} \frac{\alpha_i u_i^0}{\lambda - (\beta - 1)\alpha_i u_i^0} \, . \tag{7.3.14}$$

One can easily verify that all its roots are real and positive if $\beta > 1$. Hence, all perturbations of stationary populations u_i within the coexisting group of species are damped.

Now we can summarize the results of our analysis in the case $\beta > 1$. We have found that equations which describe the competition of agents for a common resource, in the presence of additional specialized resources, have stable steady solutions, corresponding to the stationary coexistence of the m most efficient agents, if coefficient β satisfies the inequalities

$$\Lambda_m > (\beta - 1) > \Lambda_{m-1} \, , \tag{7.3.15}$$

where the Λ_m are defined by (7.3.7). Since the set of quantities Λ_m is fixed for a given set of reproduction efficiencies k_i, conditions (7.3.15) uniquely determine the *size* m of the coexisting group as a function of β.

When $1 < \beta < (1 + \Lambda_1)$ no coexistence is possible: only the most efficient first species survives. In the interval $(1 + \Lambda_1) < \beta < (1 + \Lambda_2)$ the first two species can coexist, and so on. Finally, when $\beta > (1 + \Lambda_{N-1})$ coexistence of all N species sets in.

Hence, we see that, under the continuous variation of β, there is a sequence of discrete transitions which change the size of the coexisting group.

If we interpret (7.3.1) as a model with mutual inhibition of reproductive agents, the strength of such mutual inhibition is characterized by the inverse of β. In this interpretation the above results mean simply that the size of the coexisting group of agents increases when mutual inhibition becomes weaker; coexistence begins at $\beta = 1 + \Lambda_1$.

The case of strong inhibition ($\beta < 1$) requires a separate treatment. First of all, it can be shown that coexistence is impossible in this situation. Indeed, although a formal stationary solution (7.3.5) continues to hold in such a case, it is not furthermore stable with respect to internal perturbations. Indeed, equation (7.3.14) then takes the form

$$1 = \sum_{i=1}^{m} \frac{\alpha_i u_i^0}{\lambda + (1 - \beta)\alpha_i u_i^0} \tag{7.3.16}$$

It can be easily seen that for $m > 1$ this has negative roots λ which lead to growing perturbations.

Hence, we have to examine only single-species solutions. Suppose that only a certain jth species has a nonvanishing steady population, so that (cf. (7.3.1))

$$u_j^0 = k_j/\beta \,,$$
$$u_j^0 = 0 \quad \text{for all} \quad i \neq j \,. \tag{7.3.17}$$

This single-species solution is stable if the growth of all other species is suppressed. As follows from (7.3.1), populations of all other species are damped when

$$k_i - k_j/\beta < 0 \quad \text{for all} \quad i \neq j \,. \tag{7.3.18}$$

In effect, it is sufficient to satisfy this inequality for $i = 1$, since the first species has the highest reproduction efficiency k_1. Thus we find that, at a given β, any species with $k_j > \beta k_1$ can suppress reproduction of all other species and form a stable single-species state.

The total number of species that are endowed with such an ability depends on the value of β, i.e. on the strength of mutual inhibition. When inhibition is relatively weak, i.e. $1 > \beta > k_2/k_1$, there is only one single-species state that corresponds to the survival of the first species. For stronger inhibition, if $k_2/k_1 > \beta > k_3/k_1$, either the first or the second species survives, depending on the initial conditions. Generally, when the condition

$$k_m/k_1 > \beta > k_{m+1}/k_1 \tag{7.3.19}$$

is satisfied, there are m different stable single-species solutions ($j = 1, 2 \ldots, m$). When mutual inhibition is very strong, i.e. $k_N/k_1 > \beta$, any of N species can survive by suppressing the reproduction of all others.

Hence, we see that reproductive networks with a sufficiently high degree of mutual inhibition possess many stable stationary states, each corresponding to a particular surviving species. In mathematical terms every such stable state represents some attractive fixed point of differential equations (7.3.1). These fixed points have their attraction basins. The outcome of evolution then depends crucially on the initial conditions.

On the other hand, when inhibition is weak enough the networks possess a single attractive fixed point which corresponds to stationary coexistence of some group of species. Such a group is selected in the process of evolution independently of initial conditions (but provided that initially all species were present at least in small numbers). The same effect is observed when competition for a common resource is facilitated by bringing in specialized resources for each competing agent.

7.4 Waves of Reproduction

Above we considered networks with nonspecific interactions. However, in realistic reproductive networks, agents often interact directly with one another, either increasing or decreasing their reproduction rates. As an example, we can mention trophic chains in ecological systems. The products of a given species (or the individuals of this species themselves) can be consumed by another species which, in its turn, can provide food for yet another one, etc. It is also possible that the product (or the waste) of a particular species is poisonous for another one. Similar processes also take place, in effect, in industrial production networks.

In this section we consider a model of a reproduction network with directed interactions, given by a system of equations

$$\dot{u}_i = \alpha_i \left(k_i - \beta_i u_i - \sum_{j \neq i} J_{ij} u_j \right) u_i + \mu_i . \tag{7.4.1}$$

It differs from (7.1.12) in the last term μ_i which describes a small constant flow of population from some external depot (for instance, from an *ecological niche*, which is a largely autonomous reproduction area where competition is absent and a steady population level is maintained).

It is convenient to introduce new variables $u_i' = (\beta_i/k_i)u_i$, so that equations (7.4.1) are transformed to

$$\dot{u}_i' = \alpha_i' \left(1 - u_i' - \sum_{j \neq i} J_{ij}' u_j' \right) u_i' + \mu_i' , \tag{7.4.2}$$

where

$$\alpha_i' = \alpha_i k_i , \quad \mu_i' = (\beta_i/k_i)\mu_i , \quad J_{ij}' = (k_j/k_i\beta_j)J_{ij} . \tag{7.4.3}$$

Since only the transformed equations are studied below, we omit the primes everywhere, and write (7.4.2) simply as

$$\dot{u}_i = \alpha_i \left(1 - u_i - \sum_{j \neq i} J_{ij} u_j \right) u_i + \mu_i , \tag{7.4.2a}$$

Positive values of J_{ij} correspond to inhibitory interactions. One can easily see that if inhibition is strong, so that all coefficients J_{ij} satisfy the condition $J_{ij} > 1$, and the external sources are absent ($\mu_i = 0$), this network is multistable. It possesses a set of stable single-species states

$$u_i = 1 , \; u_j = 0 \quad \text{for} \quad j \neq i . \tag{7.4.4}$$

Hence, an arbitrary species can survive, suppressing all other species. In a more general case, when small external population sources ($1 \gg \mu_i > 0$) are present, complete extinction of other species is prevented and some small, yet nonvanishing, level of population is maintained for all of them, i. e.[2]

$$u_i \approx 1, \; u_j \approx \mu_i \alpha_i^{-1}(J_{ij} - 1)^{-1} \quad \text{for} \quad j \neq i .$$ (7.4.5)

When the condition $0 < J_{ij} < 1$ holds, inhibition is not strong enough to suppress individual reproduction of other species. Such weakly inhibitory interactions do not differ essentially, in this sense, from activatory interactions with negative values of J_{ij} and we can consider them together. Below, the connections with $J_{ij} < 1$ are called *effectively activatory*. Note that, since interactions are not symmetric ($J_{ij} \neq J_{ji}$), every connection is directed.

By indicating only the connections with $J_{ij} < 1$, we can construct a graph of effectively activatory connections for a given reproductive network (Fig. 7.1).

Let us consider some site i in the network that has only one outgoing effectively activatory connection, leading to another site k. If we start from an initial condition, where $u_i = 1$ and all other populations almost vanish, we find from (7.4.2a) that the population of species k increases with time as

$$\dot{u}_k = \alpha_k(1 - J_{ki})u_k + \mu_k$$ (7.4.6)

While u_k remains much less than unity this exponential growth has no significant reverse influence on the population species i. However, at larger values of u_k its reverse action on u_i should be taken into account.

Suppose that such reverse action is strongly inhibitory (i.e. $J_{ik} > 1$). Then species k strongly inhibits reproduction of species i, and therefore population u_i will begin to decrease when u_k is sufficiently large. This process will continue as long as u_k does not reach the steady value $u_k \approx 1$ and u_i does not decline to a very low level $u_i \approx \mu_i(J_{ik} - 1)^{-1}$.

Hence, we see that, after some interval of time, reproductive activity at site i will fall off, simultaneously raising reproductive activity at site k of the network.

The significant transition begins when population u_k approaches unity. Therefore, time interval T_i of increased reproductive activity at site i can be estimated from (7.4.6) as

$$T_i \sim \alpha_i^{-1}(1 - J_{ki})^{-1}|\ln \mu_k| .$$ (7.4.7)

On the other hand, the duration of transition, defined as the characteristic time within which both populations are comparable in their magnitude, is

$$\tau_i \sim \alpha_i^{-1}(1 - J_{ki})^{-1} .$$ (7.4.8)

Thus, when the flow of population μ_k from external sources is very small, these two time scales, i.e. transition time τ_i and activity time T_i, are sharply different. In this case a simple description of the network dynamics can be suggested. Namely, we can treat the spot of increased activity as an "excitation" which is able to move from one site in the network to another.

---

[2] Note that we measure populations u_i in certain units; the actual number of reproducing individuals of a species k may be large even if $u_k \ll 1$.

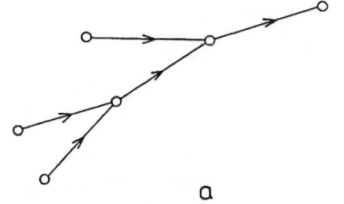

Fig. 7.1a–c. Possible graphs of effectively activatory connections: (a) with a terminal site, (b) with a simple loop, (c) with a fork vertex

a

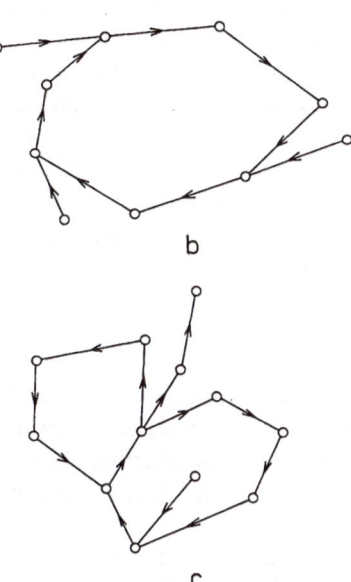

b

c

This excitation stays for a long time T_i at site i and then quickly jumps within time τ_i to the neighboring site k. If site k has again only one neighbor with an effectively activatory connection, the process is repeated and the excitation will jump after some time from site k to the next location. Obviously, such a process provides a mechanism for excitation motion through the network, i.e. for propagation of a *reproduction wave*.

The trajectory of propagation is determined by the graph of effectively activatory connections. When this graph has a vertex with an incoming connection but without any outgoing connection, as shown in Fig. 7.1a, propagation is terminated when the excitation reaches the corresponding site in the network. Then the system becomes trapped in a steady stationary state, in which this particular site is populated and populations of all other sites are vanishingly small.

It is also possible that the graph of effectively activatory connections includes a closed path (Fig. 7.1b). In this case the excitation can circulate indefinitely around such a loop, which results in stable periodic oscillations in the reproductive behavior of the system.

This phenomenon was found by *Eigen* and *Schuster* [7.2] for the reproductive chemical networks of prebiological evolution. Although mathematical models of such networks are more complicated than (7.4.1), their dynamic properties are essentially the same. In the context of biochemical models with catalytic reactions, a closed path in the graph of activatory connections is called a *hypercycle* (see [7.2]).

In more complicated graph topologies, fork sites with several outgoing connections are possible (Fig. 7.1c). When the population at such a site i is high (i. e. about unity), it triggers population growth in all neighboring sites that are directly linked to i by effectively activatory connections. So long as the populations in the neighboring sites remain small, their growth is described by the linear equation (7.4.6). However, when the populations increase further, they begin to act in a strongly inhibitory way not only backwards onto the population at the site i, but also onto each other[3].

This strong inhibition between the growing populations prevents their final steady coexistence (cf. Sect. 7.3). When these populations are sufficiently large, competition between them sets in. In the absence of population flow from the external depots, this would result in survival of one of the species and extinction of all the others. Weak external flows, described by terms μ_i in (7.4.2), prevent ultimate extinction; still the populations of other species in the final state would be vanishingly small (cf. (7.4.5)).

Hence, when an excitation arrives at a fork site with several outgoing activatory connections, it is not simultaneously transferred along all such connections. Instead it jumps to one of the connected sites.

The choice of a particular path is not unique; it is determined by the initial conditions for the competing species (i. e. by small background populations due to external sources). Therefore, in the vicinity of such vertexes, the reproductive system is very sensitive to perturbations: even small variations in the parameters can result in changes of the propagation path.

It should be emphasized that the simple description given above of reproduction waves is possible only if the background population level is very low compared with the population of a site in the "excited" state. As follows from (7.4.5), this condition holds if, for all the sites i and j linked by strongly inhibitory connections (with $J_{ij} > 1$), we have

$$\mu_i \ll \alpha_i(J_{ij} - 1) . \tag{7.4.9}$$

Furthermore the two characteristic times T_i and τ_i should be sharply different. An excitation stays long enough at each site to permit the background populations at other sites to adjust to its new position. Hence, after each quick jump the network has enough time to relax to a new quasi-stationary state.

When sources μ_i are not weak or coefficients J_{ij} are very close to unity, these conditions are violated. In this case the dynamics of a reproductive network (7.4.2) becomes much more complicated (and possibly stochastic).

Discussion of further mathematical aspects of reproductive networks with directed interactions can be found in [7.11–13].

---

[3] This graph shows only effectively activatory connections. When a connection between two sites is not indicated, it means the connection is strongly inhibitory ($J_{ij} > 1$).

7.5 Evolution and Information Processing

The analysis performed in this chapter shows that reproductive networks can demonstrate complex dynamic behavior which is highly sensitive to changes in the initial conditions and/or in the parameters of the system. Moreover, models of reproductive networks with directed interactions, studied in the previous section, have a dynamics which is similar in many respects to that of neural networks[4].

Because of the complexity in the dynamic behavior of reproductive networks, they are able to realize various functions of analog information processing. In effect, many natural reproductive systems have certain "intelligent" properties.

A living cell is essentially a network of many thousands of catalytic biochemical reactions that involve reproduction (i. e. autocatalysis), activation, and inhibition. All these reactions are strictly coordinated to produce regular and predictable spatio-temporal behavior. Moreover, cells react in a variety of different ways to possible changes in their environment by adjusting their various rates of production or even by switching from one kinetic regime to another. If we were to try to simulate numerically the reaction kinetics inside a single cell, we would find that even the best modern computers were not sufficiently powerful. Insofar as the activity of a living cell is related to information processing, we can consider it as a highly specialized natural "biocomputer". This point of view was advocated and further elaborated by *Conrad* [7.17–20] and *Liberman* [7.21].

At a macro level, market economics is another example of a complex reproductive network which is capable of performing complicated tasks of information processing. This network represents a system of tangled production chains with complex activatory (cooperation) and inhibitory (competition) interactions between the individual agents. By using its self-organization properties, such a system can efficiently optimize the costs of its final products or adjust itself to variations in consumer demand or in the environment.

This list can easily be continued. Instead of presenting further examples, we discuss below some general theoretical possibilities for analog information processing by reproductive networks.

One fundamental operation of analog information processing is pattern recognition, which is intricately related to the property of associative memory. As already noted in Sect. 1.2, this operation can be realized by any dynamical system that possesses multiple stationary attractors.

Suppose we have a reproductive network (7.3.1) with strongly inhibitory interactions between species (i. e. $\beta < k_{min}/k_{max}$). In this case the network has a set of stable stationary states that correspond to survival of one species and extinction of all the others. Depending on the initial conditions, the network evolves to a certain steady state from such set.

Let us assume now that we can map in some way the patterns which we want to recognize onto the initial populations of species in this reproductive network. Then any such pattern will trigger evolution of the network towards a particular attractive

---

[4] Functional analogies between neural networks of the brain and the reproductive system of interacting immune cells were noted by *Jerne* [7.14–16].

stationary state. In other words, all these patterns will be classified in a number of different groups, each corresponding to final survival of a particular species. If we find a suitable law of mapping, each group can be made to correspond to a definite prototype pattern, so that this reproductive network will perform the task of pattern recognition. Moreover, by changing the values of the coefficients k_i in (7.3.1) we can modify the sizes of attraction basins of stationary states, which allows us to control the required similarity with the prototype patterns.

Furthermore, one can imagine a situation where each species (which can represent, for example, a certain kind of molecules) develops its uniquely associated pattern when its population exceeds some threshold level. In this case each pattern presented for recognition (and mapped onto the initial conditions) will be not only classified, leading to enhancement in the population of a particular species, but will also produce an associated specific response in the form of a certain pattern, depending on the result of the recognition. In other words, this system will be endowed with the property of associative memory[5].

In an alternative approach, analog pattern recognition can be based on the properties of the selection process discussed in Sect. 7.2. In this case all species compete for the same common resource and competition results in survival of the fittest species and extinction of all others, independent of the initial conditions. This process is described by the models that have, for a given set of parameters, a *unique* stationary attractor which corresponds to a state where populations of all species, except for the fittest, vanish. However, when the parameters are changed this can also result in a change of attractor, i. e. some new species can become the fittest and win the competition. Hence, this dynamical system is very sensitive to changes in the parameter set. Such sensitivity can be used to realize pattern recognition, as it was noted by *Mikhailov* [7.22].

Suppose we have a dynamical system which is specified by parameters Ω_1, Ω_2, ..., Ω_N. At any given values of these parameters the system possesses a unique attractive stationary state O, so that, independent of the initial condition, its evolution always ends in this stationary state. Furthermore, when the parameters are changed, this can result in disappearance of the old attractor and emergence of a new one. Let us denote by $\{O_1, \ldots, O_M\}$ a set of all the attractive stationary states that are possible for different values of the parameters. If we consider a *parameter space*, where every point is given by a certain combination $\{\Omega_1, \Omega_2, \ldots, \Omega_N\}$ of the parameters, then each possible attractive state O, would define some region in this parameter space, i. e. the parameter basin of this attractor.

Suppose further that we map different patterns, presented for recognition, into different parameter combinations $\{\Omega_1, \Omega_2, \ldots, \Omega_N\}$, so that every such pattern is put into correspondence with some point in the parameter space. This point is associated with a certain attractor that would be reached asymptotically by the system at long times. Thus, by mapping the patterns into the parameter space and following the time evolution of our dynamic system, we can classify the patterns into groups, each corresponding to the parameter region of a particular attractor. If the dynamical

---

[5] A similar idea is used in a dynamical model of associative memory proposed by *Fuchs* and *Haken* [7.21]. However, the multistable system of their model has no direct interpretation in terms of a reproductive network.

system is engineered in such a way that every parameter region defines a class of patterns that are close to a certain prototype, this classification will result in pattern recognition. When each possible attractor represents, in its turn, some activity pattern, we obtain a system with associative memory. Note that, in contrast with the above models, the initial conditions are now insignificant since association is performed in the parameter space, and for each set of values of the parameters this system has a unique attractive fixed point.

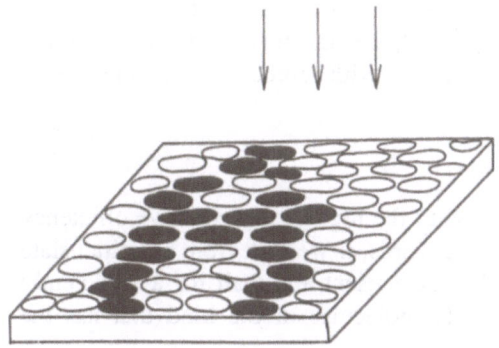

Fig. 7.2. Flat layer of cells with the projected pattern (letter 'A')

To illustrate the operation of such an associative memory, let us consider a hypothetical model with catalytic chemical reactions in a layer of cells, proposed by *Gerzon* and *Mikhailov* [7.23]. In this model (see Fig. 7.2) we have a set of cells that form a flat layer. Inside these cells certain chemical reactions go on; cell membranes are transparent for some chemicals and impenetrable for the others. The cells are immersed in a liquid medium where the reacting molecules can diffuse, thus realizing interactions between different cells. The rates of reactions are sensitive to the intensity of light that falls on a cell.

Suppose we optically project onto this flat layer some pattern $\{I_j\}$, so that a cell j is lit if $I_j = 1$ or dark if $I_j = 0$. When a cell is lit, it can support reproduction of M different kinds of molecules. However, the molecules of kind m can reproduce inside a given cell j only if it contains some specific agent. This property of a cell is quantified by a variable χ_j^m that takes value 1 when the agent is present and value 0 when it is absent. The reproduction process requires a certain substratum which is common for all kinds m of reproducing molecules. We assume that the cell membranes are transparent both for the substratum and the reproducing molecules, which can easily penetrate into the intercellular liquid. Diffusion is fast enough to ensure ideal mixing.

The kinetic equations for concentrations n_m of molecules m are

$$\dot{n}_m = \nu \left(N^{-1} \sum_j I_j \chi_j^m \right) c n_m - \gamma n_m \ . \tag{7.5.1}$$

Here c is the substratum concentration, N is the total number of cells, ν is the characteristic reproduction rate; the last term describes decay of the reproducing molecules. These equations take into account that the reproduction of molecules m inside a cell j occurs only if it is lit ($I_j = 1$) and includes a specific agent ($\chi_j^m = 1$). Hence $\sum_j I_j \chi_j^m$ is the number of cells where molecules m reproduce.

The substratum dynamics is described by equation

$$\dot{c} = -\nu \sum_m \left(N^{-1} \sum_j I_j \chi_j^m \right) n_m c - \Gamma c + S , \qquad (7.5.2)$$

where S is a constant source and Γ is the decay rate of the substratum.

Comparing this model with 7.1.3) and (7.1.4), we find that it describes competition for a common resource c between species m with reproductive efficiencies

$$s_m = \nu N^{-1} \sum_j I_j \chi_j^m . \qquad (7.5.3)$$

This process leads to selection of a species with the highest reproductive efficiency s_m and extinction of all other species. In other words, if we start from an initial state in which all kinds of reproducing molecules are present and wait for a sufficiently long time, we finally find in the system only molecules of the kind that has the highest value of s_m.

Now suppose that each distinct kind m of reproducing molecule corresponds to a certain prototype pattern. This pattern is stored in the set of values χ_j^m which specify the presence or absence of reproductive agents for the mth species in the jth cell. For simplicity let us assume that neither the prototype patterns nor the pattern being recognized are biased, i.e. conditions

$$\sum_j I_j = N/2 , \quad \sum_j \chi_j^m = N/2 \qquad (7.5.4)$$

are satisfied. Then each pattern consists of an equal number of dark and lit cells (N is the total number of cells).

As follows from (7.5.3), each reproductive efficiency s_m is proportional to the overlap between the projected pattern $\{I_j\}$ and the corresponding stored mth prototype pattern $\{\chi_j^m\}$. Hence, when some pattern $\{I_j\}$ is projected, the system of cells responds by increasing the reproduction rate of the kind of molecule associated with the nearest stored prototype pattern. In the course of time, the molecules of all other kinds will disappear and only the molecules with the highest reproductive rate will be found in the system. Therefore, by observing the time evolution of the system, one can easily determine which of the stored patterns is closest to the pattern being projected. In effect, different kinds of reproducing molecules play here the role of indicators: the presence of a particular kind m of these molecules in the final state indicates that the projected pattern is recognized as a variant of the stored mth prototype.

Further possibilities for analog information processing are provided by the waves of reproduction in the networks with directed interactions which were studied in Sect. 7.4. As we have seen, these networks might include closed loops of connected sites around which an "excitation" can propagate. When this "excitation" stays at a particular site, the reproductive activity of the corresponding species is largely enhanced. It propagates by jumps from one site in the loop to another. A network can simultaneously include several different loops, so that different routes of periodic propagation are possible and the particular route is determined by the initial conditions.

Suppose we have mapped each site in the reproductive network to some external pattern. By applying this pattern, we increase the population of the corresponding species. On the other hand, when the population of a species exceeds some threshold, the corresponding pattern is developed.

Fig. 7.3. Graph of a reproductive network with two loops (only effectively activatory connections are shown)

If we apply some pattern at the initial time moment, this creates an initial condition with the "excitation" sitting at the corresponding site. Afterwards it will start to propagate and will be attracted by the loop to whose basin the initial site belongs (Fig. 7.3). Propagation of an excitation (i. e. of a reproduction wave) around this loop will periodically increase the populations of involved species. Since each increased population develops a particular external pattern, the initial pattern will trigger a periodic sequence of such patterns, determined by a corresponding propagation loop. When there are several different such loops in the network, the choice of a particular sequence of patterns is realized by an initial condition. Hence, we see that the reproductive networks are able to store and retrieve temporal sequences.

An alternative approach can consist in mapping external patterns to the strengths J_{ij} of directed connections in a reproductive network. This allows one to control the network dynamics by application of different stationary patterns. In some cases even a small external interference is sufficient to change the dynamics qualitatively. For instance, if a graph of effectively activatory connections includes a fork vertex (see Sect. 7.4), small changes in the relative connection strengths can be enough to channel the excitation propagation from one branch of the loop to another.

Evolutionary systems with random mutations are well suited to solve optimization tasks. This property was investigated by *Boseniuk, Ebeling* and *Engel* [7.24] who considered an example where the optimization task consisted in finding the shortest route in the travelling salesman problem discussed in Sect. 6.3.

Finally, it should be noted that evolution is very intimately related to learning. As we have mentioned, natural reproductive networks are often capable of performing very "intelligent" feats of information processing. This is done in an analog manner and thus requires an appropriate internal organization of the network. To construct an artificial reproductive network for analog information processing, one should deliberately engineer its structure and the interactions between its agents. Obviously, such intentional engineering was never performed in systems which developed in a natural way in the process of evolution. Instead of that, their internal structure was formed by an interplay of random mutations and the selection process.

The simple evolutionary models discussed in Sect. 7.2 describe exceptional situations where we have only indirect competition for a common resource in a homogeneous group of species. If we consider a realistic reproductive system, we usually find that it consists of a number of specialized reproductive agents which are included in a complex network of cooperative interactions[6]. Then the reproductive efficiency of a single agent is closely dependent on the functional efficiency of the entire network, and selection operates at a macro level for the entire cooperative networks. This idea is central in *sociobiology* (see, e. g., [7.25–26]).

However, mutations are individual, i.e. they lead to the emergence of a new agent in a network. When a new agent appears, this can either improve or impair the overall behavior of the network. In other words, this can result either in some loss or in some additional profit for the entire system.

To make the evolution effective, there should exist some mechanism of "profit distribution" which allows the propagation back to an individual agent of the global consequences of a mutation, paying to an agent to the degree to which it contributed into the global gain and thus stimulating or damping its production. In this way a particular new agent will be eliminated if it does not participate in the joint operation of a network or impairs its purposeful behavior. On the other hand, if a new agent improves the overall operation of a network, it will develop its production and oust other less efficient agents.

Hence, by a process of consequent mutations and selection, controlled by a kind of back-propagation of profits, the reproductive cooperative network can gradually improve its operation until it reaches a final optimum (if it exists). Since this process is realized by the emergence of better agents and the elimination of worse ones, it leads to a gradual self-reorganization of the reproductive network during which both the topology of connections and their strengths change.

Such internal optimization of the network structure for the solution of a particular task represents *learning*, as it is defined for analog information processing devices based on distributed active systems. Hence, we can say that any evolving system gradually builds up its internal organizataion by learning the requirements of the surrounding world.

It seems very interesting to implement these principles in artificial devices for analog information processing. As shown by *Farmer, Packard* and *Perelson* [7.27], this approach is actually used in the "classifier system" that was introduced by *Holland* [7.28] for the solution of some special optimization tasks. Moreover, the effective reproductive network which arises in this application greatly resembles the natural cooperative networks of the immune system, which is also able to learn by mutation and selection. Further theoretical studies in this direction were carried out in [7.29, 30].

---

[6] Examples of such cooperative networks are multicellular biological organisms, ecological biocoenoses, human communities, etc.

References *

Chapter 1

1.1   H. Haken: *Synergetics. An Introduction* Springer Ser. Synergetics. Vol. 1 (Springer, Berlin, Heidelberg 1978)
1.2   H. Haken: *Advanced Synergetics* Springer Ser. Synergetics Vol. 20 (Springer, Berlin, Heidelberg 1987)
1.3   L. von Bertalanffy: *Problems of Life* (Watts, London 1952)
1.4   P.A. Corning: *The Synergism Hypothesis* (McGraw-Hill, New York 1983)
1.5   G. Nicolis, I. Prigogine: *Self-Organization in Nonequilibrium Systems* (Wiley, New York 1977)
1.6   L.S. Polak, A.S. Mikhailov: *Self-Organization in Nonequilibrium Physicochemical Systems* (Nauka, Moscow 1983) [in Russian]
1.7   E. Schrödinger: *What is Life?* (Cambridge Univ. Press 1944)
1.8   P. Glensdorf, I. Prigogine: *Thermodynamic Theory of Structure, Stability and Fluctuations* (Wiley, New York 1971)
1.9   H. Haken: Rev. Mod. Phys. **47**, 67 (1975)
1.10 S. Wolfram: Rev. Mod. Phys. **55**, 601 (1983)
1.11 S. Wolfram: Physica **10D**, 1–35 (1986)
1.12 M. Gardner: *Wheels, Life and Other Mathematical Amusements* (Freeman, San Francisco 1982)
1.13 F. Dyson: *Disturbing the Universe* (Harper & Row, London 1979)
1.14 H. Haken: "Pattern formation and pattern recognition – an attempt at a synthesis" in *Pattern Formation by Dynamic Systems and Pattern Recognition*, ed. by H. Haken, Springer Ser. Synergetics Vol. 5 (Springer, Berlin, Heidelberg 1979)
1.15 A.S. Mikhailov: "Engineering of dynamical systems for pattern recognition and information processing" in *Nonlinear Waves. Dynamics and Evolution*, ed. by A.V. Gaponov-Grekhov, M.I. Rabinovich (Springer, Berlin, Heidelberg 1989)

Chapter 2

2.1   F. Schlögl: Z. Phys. **253**, 147 (1982)
2.2   V.A. Vasilev, Yu.M. Romanovskii, D.S. Chernavskii, V.G. Yakhno: *Autowave Processes in Kinetic Systems* (Reidel, Dordrecht 1986)
2.3   A.G. Merzhanov, E.N. Rumanov: Usp. Fiz. Nauk **151**, 553–593 (1987)

---

* The majority of Soviet physical journals are translated into English. Below only the references to original publications are given. For convenience, we provide the table of some translated titles:

Usp. Fiz. Nauk	Sov. Phys. – Usp.
Zh. Eksp. Teor. Fiz.	Sov. Phys. – JETP
Pisma Zh. Eksp. Teor. Fiz.	Sov. Phys. – JETP Lett.
Dokl. Akad. Nauk SSSR	Sov. Phys. – Dokl.
Biofizika	Sov. Biophysics
Izv. VUZ. Radiofizika	Sov. Radiophysics
Teor. Mat. Fiz.	Sov. Phys. – Theor. Math. Phys.

2.4   A.V. Gurevich, R.G. Mints: Usp. Fiz. Nauk **142**, 61–98 (1984)
2.5   R. Luther: Z. Elektrochemie **12**, 596–600 (1906); reprinted in: J. Chem. Educ. **64**, 740 (1987)
2.6   P. Fisher: Ann. Eugenics **7**, 335–367 (1937)
2.7   Ya.B. Zeldovich, D.A. Frank-Kamenetskii: Zh. Fiz. Khim. **12**, 100–105 (1938)
2.8   Ya.B. Zeldovich, G.I. Barenblatt, V.B. Librovich, G.M. Makhviladze: *Mathematical Theory of Combustion and Explosion* (Consultants Bureau, New York 1985)
2.9   G.H. Markstein: J. Aeronaut. Sci. **18**, 199–207 (1951)
2.10   A.N. Kolmogorov, I.G. Petrovskii, N.S. Piskunov: Bull. Mosk. Gos. Univ., Sekt. Matematika i Mekhanika **1**, 1–26 (1937)
2.11   T.S. Akhromeyeva, S.P. Kurdyumov, G.G. Malinetskii, A.A. Samarskii: Phys. Rep. **176**, 189–370 (1989)

Chapter 3

3.1   B.P. Belousov: "Periodically acting reaction and its mechanisms", in *Sbornik Referatov po Radiatsionnoy Meditsine* (Medgiz, Moscow 1959) p. 145
3.2   A.I. Zhabotinskii: Dokl. Akad. Nauk SSSR **157**, 392–395 (1964)
3.3   A.I. Zhabotinskii: *Concentration Self-Oscillations* (Nauka, Moscow 1974)
3.4   J.J. Tyson: *The Belousov-Zhabotinsky Reaction*, Lect. Notes in Biomath., Vol. 10 (Springer, Berlin, Heidelberg 1976)
3.5   J.J. Tyson, P.C. Fife: J. Chem. Phys. **73**, 2224–2236 (1980)
3.6   R.J. Field, E. Köros, R. Noyes: J. Am. Chem. Soc. **94**, 8649–8664 (1972)
3.7   A.C. Scott: Rev. Mod. Phys. **47**, 47–92 (1975)
3.8   V.A. Vasilev, Yu.M. Romanovskii, D.S. Chernavskii, V.G. Yakhno: *Autowave Processes in Kinetic Systems* (Reidel, Dordrecht 1986)
3.9   P. Ortoleva, J. Ross: J. Chem. Phys. **63**, 3398–3431 (1975)
3.10   L.A. Ostrovskii, V.G. Yakhno: Biofizika **20**, 489–496 (1975)
3.11   R.C. Casten, H. Cohen, P.A. Lagerstrom: Quart. Appl. Math. **32**, 365–402 (1975)
3.12   J. Murray: *Lectures on Nonlinear Differential Equation Models in Biology* (Clarendon Press, Oxford 1977)
3.13   A.S. Mikhailov, V.I. Krinsky: Physica **9D**, 346–376 (1983)
3.14   J.A. Feroe: Biophys. J. **21**, 103–110 (1978)
3.15   K. Maginu: J. Math. Biol. **6**, 49–57 (1978)
3.16   J. Rinzel, J.B. Keller: Biophys. J. **13**, 1313–1337 (1973)
3.17   L.S. Polak, A.S. Mikhailov: *Self-Organization in Nonequilibrium Physicochemical Systems* (Nauka, Moscow 1983)
3.18   V.S. Zykov: Biofizika **25**, 888–892 (1980)
3.19   V.S. Zykov: *Simulation of Wave Processes in Excitable Media* (Nauka, Moscow, 1984) [English transl.: Manchester Univ. Press, 1987]
3.20   N. Wiener, A. Rosenblueth: Arch. Inst. Cardiol. Mex. **16**, 205–265 (1946)
3.21   A. Rosenblueth: Amer. J. Physiol. **194**, 491 (1958)
3.22   O. Selfridge: Arch. Inst. Cardiol. Mex. **53**, 113 (1948)
3.23   I.S. Balakhovskii: Biofizika **10**, 1063–1069 (1965)
3.24   V.I. Krinsky: "Fibrillation in excitable media" in *Problemy Kibernetiki*, Vol. 20 (Nauka, Moscow 1968) pp. 59–73
3.25   L.V. Reshodko, J. Bureš: Biol. Cybern. **18**, 181–189 (1975)
3.26   B.F. Madore, W.L. Freedman: Science **222**, 615–617 (1983)
3.27   A.T. Winfree, E.M. Winfree, H. Seifert: Physica **17D**, 109–121 (1985)
3.28   A.S. Mikhailov, V.S. Zykov: Dokl. Akad. Nauk SSSR **286**, 341–344 (1986)
3.29   B. Dorjsurengiyn: Diploma Thesis (Dept. Physics, Moscow State Univ., 1988)
3.30   A.S. Mikhailov: Priroda **3**, 15–26 (1987)
3.31   L. Kuhnert: Nature **319**, 393 (1986)
3.32   A.M. Zhabotinskii, A.N. Zaikin: Nature **225**, 535–538 (1970)

3.33 K.I. Agladze, V.I. Krinsky: "On the mechanism of target pattern formation in the distributed Belousov-Zhabotinsky system" in *Self-Organization: Autowaves and Structures far from Equilibrium*, ed. by V.I. Krinsky (Springer, Berlin, Heidelberg 1984) pp. 147–149

3.34 A.M. Zhabotinskii, A.N. Zaikin: "Spatial effects in a self-oscillatory chemical system" in *Kolebatelnye Protsessy v Biologicheskikh i Khimicheskikh Sistemakh* [Oscillatory Processes in Biological and Chemical Systems], Vol. 2 (Biological Research Center Publications, Pushchino 1971) pp. 314–317

3.35 A.M. Zhabotinskii, A.N. Zaikin: J. Theor. Biol. **40**, 45–61 (1973)

3.36 A.T. Winfree: Science **175**, 634–640 (1972)

3.37 G.R. Ivanitskii, V.I. Krinsky, A.N. Zaikin, A.M. Zhabotinskii: Sov. Sci. Rev. **2D**, 280–350 (1980)

3.38 K.I. Agladze, V.I. Krinsky: Nature **296**, 424–426 (1982)

3.39 S.C. Müller, T. Plesser, B. Hess: Science **230**, 661 (1985)

3.40 S.L. Müller, T. Plesser, B. Hess: Naturwissenschaften **73**, 165–179 (1986)

3.41 S.L. Müller, T. Plesser, B. Hess: Physica **24D**, 71–86, 87–96 (1987)

3.42 G.K. Korn, T.M. Korn: *Mathematical Handbook for Scientists and Engineers* (McGraw-Hill, New York 1961)

3.43 A.M. Pertsov, E.A. Ermakova, A.V. Panfilov: Physica **14D**, 117–124 (1984)

3.44 A.S. Mikhailov: "A theory of spiral waves" in *Self-Organization: Autowaves and Structures far from Equilibrium* ed. by V.I. Krinsky (Springer, Berlin, Heidelberg 1984) pp. 92–95

3.45 P.K. Brazhnik, V.A. Davydov, V.S. Zykov, A.S. Mikhailov: Zh. Eksp. Teor. Fiz. **93**, 1725–1736 (1987)

3.46 V.A. Davydov, A.S. Mikhailov, V.S. Zykov: "Kinematical theory of autowave patterns in excitable media", in *Nonlinear Waves in Active Media*, ed. by Yu. Engelbrecht (Springer, Berlin, Heidelberg 1989) pp. 38–51

3.47 W.K. Burton, N. Cabrera, F.C. Frank: Phil. Trans. Roy. Soc. **243**, 299–312 (1951)

3.48 V.S. Zykov: Biofizika **25**, 319–322 (1980)

3.49 V.S. Zykov, O.L. Morozova: Biofizika **25**, 1071–1076 (1980)

3.50 P.K. Brazhnik, V.A. Davydov, A.S. Mikhailov: Teor. Mat. Fiz. **74**, 440–447 (1987)

3.51 V.A. Davydov, A. S. Mikhailov: "Spiral waves in distributed active media" in *Nonlinear Waves. Patterns and Bifurcations*, ed. by A.V. Gaponov-Grekhov, M.I. Rabinovich (Nauka, Moscow, 1987) pp. 261–279

3.52 V.S. Zykov: "Investigations of some properties of self-sustained activity in excitable medium" in *Controlling Complex Systems*, ed. By Ya.A. Tsypkin (Nauka, Moscow, 1975) pp. 59–66

3.53 R.C. Brower, D.A. Kessler, J. Koplik, H. Levine: Phys. Rev. Lett. **29**, 1335–1342 (1984)

3.54 E. Meron, P. Pelcé: Phys. Rev. Lett. **60**, 1880–1883 (1988)

3.55 P.K. Brazhnik, V.A. Davydov, A.S. Mikhailov: "Spiral waves and vortex rings in combustion with subsequent recovering of the initial properties of the medium" in *Kinetics and Combustion*, Proc. VIII Soviet Symp. on Combustion and Explosion, Tashkent, 1986 (Inst. Chem. Phys., Chernogolovka 1986) pp. 39–43

3.56 J.P. Keener, J.J. Tyson: Physica **21D**, 300–324 (1986)

3.57 J.P. Keener: SIAM J. Appl. Math. **46**, 1039–1056 (1986)

3.58 J.J. Tyson, J.P. Keener: Physica **29D**, 215–222 (1987)

3.59 J.J. Tyson, J.P. Keener: Physica **32D**, 327–361 (1988)

3.60 P.K. Brazhnik, V.A. Davydov, V.S. Zykov, A.S. Mikhailov: Izv. VUZ. Radiofizika **31**, 574–584 (1988)

3.61 K.I. Agladze, V.A. Davydov, A.S. Mikhailov: Pisma Zh. Eksp. Teor. Fiz. **45**, 601–603 (1987)

3.62 V.A. Davydov, V.S. Zykov: Zh. Eksp. Teor. Fiz. **95**, 139–148 (1989)

3.63 A.S. Mikhailov: "Kinematics of wave patterns in excitable media" in *Nonlinear Wave Processes in Excitable Media*, ed. by A.V. Holden, M. Markus, H.G. Othmer (Plenum Press, New York 1990) in press

3.64 W. Jahnke, W.E. Skeggs, A.T. Winfree: J. Phys. Chem. **95**, 740–749 (1989)

3.65 S.C. Müller, Th. Plesser: "Dynamics of Spiral Centers in the Ferroin-Catalyzed Belousov-Zhabotinskii Reaction" in *Nonlinear Wave Processes in Excitable Media*, ed. by A.V. Holden, M. Markus, H.G. Othmer (Plenum Press, New York 1990) in press

3.66   V.S. Zykov: Biofizika **32**, 337–430 (1987)

3.67   V.S. Zykov, O.L. Morozova: "Kinematic method of investigation of stability of spiral autowaves" (Preprint, Inst. of Control Science, Moscow 1988); J. Nonlinear Biol. **1** (1990), in press

3.68   A.T. Winfree, S.H. Strogatz: Physica **9D**, 35–49, 65–80, 333–345 (1983)

3.69   A.T. Winfree, S.H. Strogatz: Physica **13D**, 221–233 (1984)

3.70   Ya.B. Zeldovich, B.A. Malomed: Dokl. Akad. Nauk SSSR **254**, 92–94 (1980)

3.71   B.J. Welsh, J. Gomatam, A.E. Burgess: Nature **304**, 611–614 (1983)

3.72   A.B. Medvinsky, A.V. Panfilov, A.M. Pertsov: "Properties of rotating waves in three dimensions" in *Self-Organization: Autowaves and Structures far from Equilibrium*, ed. by V.I. Krinsky (Springer, Berlin, Heidelberg 1984) pp. 195–199

3.73   A.V. Panfilov, A.N. Rudenko: Physica **28D**, 215–218 (1987)

3.74   A.V. Panfilov, A.T. Winfree: Physica **17D**, 323–330 (1985)

3.75   A.V. Panfilov, R.R. Aliev, A.V. Mushinsky: Physica **36D**, 181–186 (1989)

3.76   P.K. Brazhnik: *Geometric Methods in the Theory of Autowave Patterns* (Ph.D. Thesis, Dept. Physics, Moscow State Univ. 1988)

3.77   N.V. Khrustova: Diploma Thesis (Dept. Physics, Moscow State Univ. 1989)

3.78   J.P. Keener: Physica **31D**, 269–276 (1988)

3.79   A.Yu. Abramychev, V.A. Davydov, A.S. Mikhailov: Biofizika **35**, 100–105 (1989)

3.80   P.K. Brazhnik, V.A. Davydov, V.S. Zykov, A.S. Mikhailov: "Dynamics of three-dimensional autowave patterns" in *Proc. II Soviet Conf. on Mathematical and Computational Methods in Biology* (Biological Research Center Publications, Pushchino 1987) pp. 118–119

3.81   P.K. Brazhnik, V.A. Davydov, A.S. Mikhailov: Izv. VUZ. Radiofizika **32**, 289–293 (1989)

3.82   A.V. Panfilov, A.N. Rudenko, A.M. Pertsov: "Twisted scroll waves in three-dimensional active media" in *Self-Organization: Autowaves and Structures far from Equilibrium*, ed. by V.I. Krinsky (Springer, Berlin, Heidelberg 1984) pp. 103–105

3.83   A.S. Mikhailov, A.V. Panfilov, A.N. Rudenko: Phys. Lett. **109A**, 246–250 (1985)

3.84   V.N. Biktashev: Physica **36D**, 167–172 (1989)

3.85   P.J. Nandapurkar, A.T. Winfree: Physica **29D**, 69–83 (1987)

3.86   M. Gerhardt, H. Schuster, J.J. Tyson: Science (1990), in press

3.87   M. Markus, B. Hess: Nature (1990), in press

3.88   G.S. Skinner, H.L. Swinney: Physica D (1990), in press

Chapter 4

4.1   G.B. Whitham: J. Fluid Mech. **44**, 373 (1970)

4.2   Y. Kuramoto: *Chemical Oscillations, Waves, and Turbulence* (Springer, Berlin, Heidelberg 1984)

4.3   L.N. Howard, N. Kopell: Stud. Appl. Math. **56**, 95–146 (1977)

4.4   N. Kopell, L.N. Howard: Stud. Appl. Math. **64**, 1–56 (1981)

4.5   Y. Kuramoto: Prog. Theor. Phys. Suppl. **64**, 346–362 (1978)

4.6   G.I. Sivashinsky: Ann. Rev. Fluid Mech. **15**, 179–199 (1983)

4.7   M.C. Cross: "Theoretical methods in pattern formation in physics, chemistry and biology" in *Far from Equilibrium Phase Transitions*, ed. by L. Garrido (Springer, Berlin, Heidelberg 1988) pp. 45–74

4.8   Y. Kuramoto, T. Tsuzuki: Prog. Theor. Phys. **55**, 356–369 (1976)

4.9   E.M. Lifshitz, L.P. Pitaevskii: *Course of Theoretical Physics. Vol.10. Physical Kinetics* (Nauka, Moscow 1979) Chap.12

4.10   A.S. Mikhailov, I.V. Uporov: Dokl. Akad. Nauk SSSR **249**, 733–736 (1979)

4.11   Ya.B. Zeldovich, B.A. Malomed: Dokl. Akad. Nauk SSSR **254**, 92–94 (1980)

4.12   Y. Kuramoto: Prog. Theor. Phys. **63**, 1885–1895 (1980)

4.13   N. Kopell, L.N. Howard: Science **180**, 1171–1173 (1973)

4.14   P. Ortoleva, J. Ross: J. Chem. Phys. **58**, 5673 (1973)

4.15   J.C. Neu: SIAM J. Appl. Math. **36**, 509 (1979)

4.16   J.C. Neu: SIAM J. Appl. Math. **38**, 305 (1980)

4.17   B.A. Malomed: Z. Phys. B **55**, 241–248, 249–256 (1984)
4.18   A.N. Zaikin, A.M. Zhabotinskii: Nature **225**, 535–538 (1970)
4.19   K.I. Agladze, V.I. Krinsky: "On the mechanism of target pattern formation in the distributed Belousov-Zhabotinsky system" in *Self-Organization: Autowaves and Structures far from Equilibrium*, ed. by V.I. Krinsky (Springer, Berlin, Heidelberg 1984) pp. 147–149
4.20   J.J. Tyson, P.C. Fife: J. Chem. Phys. **73**, 2224–2236 (1980)
4.21   J.J. Tyson: J. Chim. Physique **84**, 1359–1365 (1987)
4.22   Sh. Bose, Su. Bose, P. Ortoleva: J. Chem. Phys. **72**, 4258 (1980)
4.23   V.G. Yakhno: Biofizika **20**, 669–675 (1975)
4.24   V.A. Vasilev, Yu.M. Romanovskii, D.S. Chernavskii, V.G. Yakhno: *Autowave Processes in Kinetic Systems* (Reidel, Dordrecht 1986) Chap. 5
4.25   G.T. Gurija, M.A. Livshits: Phys. Lett. **97A**, 175–177 (1983)
4.26   V.A. Vasilev, M.S. Polyakova: Vestnik MGU. Ser. Fizika **16**, 99–104 (1975)
4.27   L.D. Landau, E.M. Lifshitz: *Course of Theoretical Physics. Vol. 3. Quantum Mechanics* (Pergamon, Oxford 1977)
4.28   A.S. Mikhailov, A. Engel: Phys. Lett. **117A** 257–260 (1986)
4.29   D.S. Cohen, J.C. Neu, R.R. Rosales: SIAM J. Appl. Math. **35**, 536–549 (1978)
4.30   T. Erneux, M. Herchkowitz-Kaufman: Bull. Math. Biol. **41**, 767–782 (1979)
4.31   J.M. Greenberg: SIAM J. Appl. Math. **39**, 301–309 (1980)
4.32   J.M. Greenberg: Adv. Appl. Math. **2**, 450 (1981)
4.33   Ya.B. Zeldovich, B.A. Malomed: Izv. VUZ. Radiofizika **25**, 591–618 (1982)
4.34   P.S. Hagan: SIAM J. Appl. Math. **42**, 762–781 (1982)
4.35   B.A. Malomed: Dokl. Akad. Nauk SSSR **291**, 327–332 (1986)
4.37   S. Koga: Prog. Theor. Phys. **67**, 164 (1982)
4.38   S. Koga: Prog. Theor. Phys. **67**, 454–463 (1982)
4.39   I.S. Aranson, M.I. Rabinovich: J. Phys. **22A**, (1989)
4.40   I.S. Aranson, M.I. Rabinovich: Izv. VUZ. Radiofizika **29**, 1514–1517 (1986)
4.41   Bodenschatz, A. Weber, L. Kramer: Dynamics and pattern of spiral waves and defects in travelling waves" in *Nonlinear Wave Processes in Excitable Media*, ed. by A.V. Holden, M. Markus, H.G. Othmer (Plenum Press, New York 1990) in press
4.42   C. Elphick, E. Meron: *Localized and Extended Patterns in Reactive Media*, Proc. IMA Workshop on Patterns and Dynamics in Reactive Media, Minneapolis (1989)

Chapter 5

5.1   V.V. Barelko, V.M. Beibutyan, Yu.V. Volodin, Ya.B. Zeldovich: Dokl. Akad. Nauk SSSR **257**, 339–344 (1981)
5.2   A.M. Turing: Phil. Trans. Roy. Soc. **237**, 37–72 (1952)
5.3   I. Prigogine, R. Lefever: J. Chem. Phys. **48**, 1695 (1968)
5.4   I. Prigogine, R. Lefever: J. Chem. Phys. **49**, 283–292 (1968)
5.5   G. Nicolis, I. Prigogine: *Self-Organization in Nonequilibrium Systems* (Wiley, New York 1977)
5.6   V.A. Vasilev, Yu.M. Romanovskii, D.S. Chernavskii, V.G. Yakhno: *Autowave Processes in Kinetic Systems* (Reidel, Dordrecht 1986)
5.7   B.N. Belintsev: Usp. Fiz. Nauk: **141**, 55–101 (1983)
5.8   B.S. Kerner, V.V. Osipov: Sov. Phys. Usp. **32**, 101–138 (1989)
5.9   B.S. Kerner, V.V. Osipov: Zh. Eksp. Teor. Fiz. **74**, 1675–1697 (1978)
5.10   B.S. Kerner, V.V. Osipov: Zh. Eksp. Teor. Fiz. **79**, 2218–2238 (1979)
5.11   B.S. Kerner, V.V. Osipov: "Autosolitons in active systems with diffusion" in *Self-Organization by Nonlinear Irreversible Processes* ed. by W. Ebeling and H. Ulbricht (Springer, Berlin, Heidelberg 1986) pp. 118–127

5.12 B.S. Kerner, V.I. Krinsky, V.V. Osipov: "Structures in models of morphogenesis" in *Thermody-namics and Pattern Formation in Biology* (de Gruyter, Berlin 1988) pp. 265–320
5.13 B.S. Kerner, V.V. Osipov: Zh. Eksp. Teor. Fiz. **83**, 2201–2214 (1982)
5.14 B.S. Kerner, V.V. Osipov: Dokl. Akad. Nauk SSSR **270**, 1104–1108 (1983)

Chapter 6

6.1   W.C. McCulloch, W. Pitts: Bull. Math. Biophys. **5**, 115–133 (1943)
6.2   M. Conrad: BioSystems **5**, 1–14 (1972)
6.3   M. Conrad: Comm. ACM **28**, 464–480 (1985)
6.4   J.J. Hopfield: Proc. Nat. Acad. Sci. USA **79**, 2554–2558 (1982)
6.5   D.C. Hebb: *The Organization of Behavior: A Neuropsycological Theory* (Wiley, New York 1957)
6.6   D.J. Amit, H. Gutfreund, H. Sompolinsky: Ann. Phys. (New York) **173**, 30–67 (1987)
6.7   T. Kohonen: *Self-Organization and Associative Memory*, Springer Ser. Information Sci. Vol. 8 (Springer, Berlin, Heidelberg 1988)
6.8   L. Personnaz, I. Guyon, G. Dreyfus: Phys. Rev. A**34**, 4217 (1987)
6.9   S. Diederich, M. Opper: Phys. Rev. Lett. **50**, 949 (1987)
6.10 W. Kinzel: Physica Scripta **39**, 171–194 (1988)
6.11 E. Gardner: J. Phys. A**21**, 256–270 (1988)
6.12 E. Gardner, B. Derrida: J. Phys. A**21**, 271 (1988)
6.13 E. Gardner, N. Stroud, D.J. Wallace: Edinburgh Preprint 87/394
6.14 W. Krauth, M. Meyard: J. Phys. A**20**, L745–L752 (1987)
6.15 G. Poppel, U. Krey: Europhys. Lett. **4**, 979 (1987)
6.16 M. Opper: Phys. Rev. Lett. **22**, 235 (1988)
6.17 H. Sompolinsky: Phys. Rev. A**34**, 2571–2574 (1986)
6.18 J.L. van Hemmen: Phys. Rev. A**36**, 1959 (1987)
6.19 B. Derrida, E. Gardner, A. Zippelius: Europhys. Lett. **4**, 167 (1987)
6.20 M.V. Tsodyks, M.V. Feigelman: Europhys. Lett. **6**, 101, 1988
6.21 J.J. Hopfield, D.W. Tank: "Collective computation with continuous variables", in *Disordered Systems and Biological Organization*, ed. by E. Bienenstock et al. (Springer, Berlin, Heidelberg 1986)
6.22 J.S. Denker: Physica **22D**, 216–232 (1986)
6.23 M.R. Garey, D.S. Johnson: *Computers and Intractability* (Freeman, San Francisco 1979)
6.24 S. Kirkpatrick, C.D. Gelatt, M.P. Vecchi: Science **220**, 671 (1983)
6.25 N. Metropolis, A.W. Rosenblueth, M.N. Rosenblueth, A.H. Teller: J. Chem. Phys. **6**, 1087–1092 (1953)
6.26 J.J. Hopfield, D.W. Tank: Biol. Cybern. **52**, 141–152 (1985)
6.27 R. Durbin, D.J. Willshaw: Nature **326**, 689–691 (1987)
6.28 F. Rosenblatt: *Principles of Neurodynamics* (Spartan Books, New York 1959)
6.29 D.E. Rumelhart, J.L. McClelland: "On learning the past tenses of English verbs", in *Parallel Distributed Processing*, Vol. 2, ed. by D.E. Rumelhart et. al. (MIT Press, Cambridge, MA 1986), pp. 216–271
6.30 M. Minsky, S. Papert: *Perceptrons* (MIT Press, Cambridge, MA 1969)
6.31 D.E. Rumelhart, G.E. Hinton, R.J. Williams: Nature **323**, 533–536 (1986)
6.32 D.E. Rumelhart, G.E. Hinton, R.J. Williams: "Learning internal representations by error propagation", in *Parallel Distributed Processing*, Vol. 1, ed. by D.E. Rumelhart et. al. (MIT Press, Cambridge, MA 1986), pp. 318–362
6.33 Y. Le Cun: "A learning scheme for asymmetric threshold networks", in Proc. Int. Conf. Cognitiva '85 (Paris, 1985) pp. 599–604
6.34 T.J. Sejnowski, P.K. Kienker, G.E. Hinton: Physica **22D**, 260–275 (1986)
6.35 R.P. Gorman, T.J. Sejnowski: Neural Networks **1**, 75–89 (1988)
6.36 S. Lehky, T.J. Sejnowski: Nature **333**, 452–454 (1988)
6.37 T.J. Sejnowski, C.R. Rosenberg: Complex Systems **1**, 145–168 (1987)

6.38 N. Quian, T.J. Sejnowski: J. Molec. Biol. **202**, 865–884 (1988)
6.39 G.E. Hinton, T.J. Sejnowski: "Optimal perceptual inference", in *Proceedings of the IEEE Computer Society Conference on Computer Vision and Pattern Recognition* (Washington, 1983) pp. 448–453
6.40 G.E. Hinton, T.J. Sejnowski: "Learning and relearning in Boltzmann machines", in *Parallel Distributed Processing*, Vol. 1, ed. by D.E. Rumelhart et al. (MIT Press, Cambridge, MA 1986) pp. 282–317
6.41 R.W. Prager, T.D. Harrison, F. Fallside: Computer Speech and Language **1**, 3–27 (1986)
6.42 P.K. Kienker, T.J. Sejnowski, G.E. Hinton, L.E. Schumacher: Perception **15**, 197–216 (1986)
6.43 D.H. Acley, G.E. Hinton, T.J. Sejnovski: Cognitive Science **9**, 147–169 (1985)
6.44 S. Amari: IEEE Trans. **C-21**, 1197–1206 (1972)
6.45 D. Kleinfield: Proc. Natl. Acad. Sci. USA **83**, 9469–9473 (1986)
6.46 L. Personnaz, I. Guyon, G. Dreyfus: Phys. Rev. **A34**, 4217 (1986)
6.47 H. Sompolinsky, I. Kanter: Phys. Rev. Lett. **57**, 2861–2864 (1986)
6.48 A. Herz, B. Sulzer, R. Kuhn, J.L. van Hemmen: Europhys. Lett. **7**, 663–669 (1988)
6.49 I. Guyon, L. Personnaz, J.P. Nadal, G. Dreyfus: Phys. Rev. **A38**, 6365–6372 (1988)
6.50 J. Buhmann, K. Schulten: Europhys. Lett. **4**, 1205–1209 (1987)
6.51 H. Rieger, M. Schreckenberg, J. Zittarz: Z. Phys. **B72**, 523 (1988)
6.52 A.S. Mikhailov, I.V. Mitkov, N.A. Sveshnikov: J. Nonlinear Biol. **1** (1990), in press
6.53 A.S. Mikhailov, I.V. Mitkov, N.A. Sveshnikov: BioSystems **23**, 291–295 (1990)

Chapter 7

7.1   M. Eigen, Naturwissenschaften **58**, 465 (1971)
7.2   M. Eigen, P. Schuster: *The Hypercycle* (Springer, Berlin, Heidelberg 1979)
7.3   J. Schumpeter: *The Theory of Economic Development* (Harvard Univ. Press, Cambridge 1934)
7.4   A.A. Alchian: J. Political Economy **58**, 211–222 (1951)
7.5   R.R. Nelson, S.G. Winter: *An Evolutionary Theory of Economic Change* (Harvard Univ. Press, Cambridge 1982)
7.6   M.A. Jimenez Montano, W. Ebeling: Collective Phenomena **3**, 107–114 (1980)
7.7   G. Silverberg: "Modelling economic dynamics and technical change: mathematical approaches to self-organization and evolution" in *Technical Change and Economic Theory*, ed. by G. Dosi et al. (Pinter, New York 1988) pp. 531–559
7.8   P. Schuster: Physica **22D**, 100–119 (1986)
7.9   W. Ebeling, R. Feistel: *Physik der Selbstorganisation und Evolution* (Akademie-Verlag, Berlin 1982)
7.10 L.D. Landau, E.M. Lifshitz: *Course of Theoretical Physics. Vol. 3. Quantum Mechanics* (Pergamon, Oxford 1977) S38
7.11 P. Schuster, K. Sigmund, R. Wolff: Bull. Math. Biol. **38**, 282 (1980)
7.12 J. Hofbauer, P. Schuster, K. Sigmund: J. Math. Biol. **11**, 155 (1981)
7.13 P.E. Phillipson, P. Schuster: J. Chem. Phys. **79**, 3807 (1983)
7.14 N.K. Jerne: Sci. Amer. No. 7, 52–60 (1973)
7.15 N.K. Jerne: Science **229**, 1057–1059 (1985)
7.16 N.K. Jerne: "Structural analogies between the immune system and the nervous system" in *Stability and Origin of Biological Information*, ed. by I.R. Miller (Wiley, New York 1975) pp. 201–204
7.17 M. Conrad: BioSystems **5**, 1–14 (1972)
7.18 M. Conrad: J. Neurosci. Res. **2**, 233–254 (1976)
7.19 M. Conrad: Comm. ACM **28**, 464–480 (1985)
7.20 E.A. Liberman: BioSystems **11**, 111–124 (1979)
7.21 A. Fuchs, H. Haken: "Computer simulations of pattern recognition as a dynamical process of a synergetic system" in *Neural and Synergetic Computers*, ed. by H. Haken, Springer Ser. Synergetics Vol. 42 (Springer, Berlin, Heidelberg 1988) pp. 16–28
7.22 A.S. Mikhailov: "Engineering of dynamical systems for pattern recognition and information processing" in *Nonlinear Waves. Dynamics and Evolution*, ed. by A.V. Gaponov-Grekhov, M.I. Rabinovich (Springer, Berlin, Heidelberg 1989)

7.23   S.A. Gerzon, A.S. Mikhailov: Dokl. Akad. Nauk SSSR **291**, 228–230 (1986)

7.24   T. Boseniuk, W. Ebeling, A. Engel: Phys. Lett. **125A**, 307–311 (1987)

7.25   R.D. Alexander: *Darwinism and Human Affairs* (Univ. Washington Press, Seattle 1979)

7.26   J.H. Crook: *Evolution of Human Consciousness* (Clarendon Press, Oxford 1980)

7.27   J.D. Farmer, N.H. Packard, A.S. Perelson: Physica **22D**, 187–204 (1986)

7.28   J.H. Holland: Physica **22D**, 307–317 (1986)

7.29   G. Weisbuch, H. Atlan, J. Phys. **A21**, L189–192 (1988)

7.30   U. Behn, J.L. van Hemmen: J. Stat. Phys. **56**, 533–545 (1989)

Subject Index

Springer Series in Synergetics

Editor: Hermann Haken

Synergetics, an interdisciplinary field of research, is concerned with the cooperation of individual parts of a system that produces macroscopic spatial, temporal or functional structures. It deals with deterministic as well as stochastic processes.